The Aqua Group Guide to Procurement, Tendering & Contract Administration

The Aqua Group Guide to Procurement, Tendering & Contract Administration

Edited and updated by

Mark Hackett, Ian Robinson & Gary Statham
Davis Langdon LLP

Blackwell Publishing Ltd editorial offices:
Blackwell Publishing Ltd, 9600 Garsington Road, Oxford OX4 2DQ, UK
Tel: +44 (0)1865 776868
Blackwell Publishing Inc., 350 Main Street, Malden, MA 02148-5020, USA
Tel: +1 781 388 8250
Blackwell Publishing Asia Pty Ltd, 550 Swanston Street, Carlton, Victoria 3053, Australia
Tel: +61 (0)3 8359 1011

This edition published 2007 by Blackwell Publishing Ltd
3 2007
Revised and updated from *Tenders and Contracts for Building* (third edition, © The Aqua Group 1975, 1982, 1990, 1999) and *Pre-contract Practice and Contract Administration* (© 2003 The Aqua Group and Blackwell Science Ltd, a Blackwell Publishing company)

Tenders and Contracts for Building first published 1975 by Crosby Lockwood Staples as *Which Builder?*; published 1982 by Granada Publishing as *Tenders and Contracts for Building*; reprinted 1986 by Collins Professional and Technical Books; second edition published 1990 by Blackwell Scientific Publications; Third edition published 1999 by Blackwell Science, a Blackwell Publishing company

Pre-contract Practice and Contract Administration published 2003 by Blackwell Science, a Blackwell Publishing company; revised and updated from *Pre-contract Practice for the Building Team* (eighth edition, © The Aqua Group 1960–1992) and *Contract Administration for the Building Team* (eighth edition, © The Aqua Group 1965–1996)

ISBN 978-1-4051-3198-8

Library of Congress Cataloging-in-Publication Data
The Aqua Group guide to procurement, tendering and contract administration/edited and updated by Mark Hackett, Ian Robinson & Gary Statham.
p. cm.
Based on the publications Pre-contract practice and Tenders and contracts for buildings.
Includes index.
ISBN-13: 978-1-4051-3198-8 (pbk. : alk. paper)
ISBN-10: 1-4051-3198-5 (pbk. : alk.paper)
1. Buildings – Specifications.
2. Construction contracts – Great Britain.
3. Letting of contracts – Great Britain.
4. Construction industry – Management.
I. Hackett, Mark, 1962– II. Robinson, Ian, 1963– III. Statham, Gary. IV. Aqua Group. V. Title: Guide to procurement, tendering and contract administration.
TH425.A678 2006
692–dc22
2006012266

A catalogue record for this title is available from the British Library

Set in 10/13 pt Palatino by Graphicraft Limited, Hong Kong
Printed and bound in Singapore by Markono Print Media Pte Ltd

The publisher's policy is to use permanent paper from mills that operate a sustainable forestry policy, and which has been manufactured from pulp processed using acid-free and elementary chlorine-free practices. Furthermore, the publisher ensures that the text paper and cover board used have met acceptable environmental accreditation standards.

For further information on Blackwell Publishing, visit our website: www.blackwellpublishing.com/construction

Contents

Foreword

Construction industry procurement has, in recent times, undergone significant change and this trend is predicted to accelerate over the next decade. Up until now, many of the industry's processes, roles and responsibilities have only changed incrementally from their traditional routes. In order for the student to understand how construction projects are procured, it is necessary to appreciate the myriad processes undertaken from the employer's initial brief through to project completion. Future industry professionals should seek to gain a breadth of knowledge covering the whole construction process, the procurement systems available and the documents establishing, governing and regulating the project. Without gaining such knowledge, the construction industry will continue to struggle in its efforts to overcome its well publicised negative characteristics.

Effective continuous improvement requires the focus of attention on industry processes and the measurement of successful outcomes which should, in turn, lead to further improvement in subsequent projects. The construction process, on the whole, requires that many professions come together as part of an integrated team that is capable of successfully delivering a project. It is the harnessing of these integrated processes and the management thereof which will ultimately determine whether an employer's expectations are met in respect of time, cost and quality.

The *Aqua Group Guide to Procurement, Tendering & Contract Administration* provides a comprehensive account of the systems, procedures, documents and industry practices encountered during construction projects. The guide also captures and contains the level of understanding that one is expected to have acquired in order to pass professional examinations and, therefore, provides a good grounding to a successful career in the construction industry.

Rob Smith
Senior Partner
Davis Langdon LLP

Introduction

For two of us, the Aqua Group Publications formed part of the reading material for our degree courses some quarter of a century ago. We always found the Aqua Books to be comprehensive, informative and readily understood. It is our hope that, as editors for the new combined title of *Procurement, Tendering & Contract Administration*, we have maintained the Aqua Group's tradition which dates back to 1960.

The combined title draws together the Aqua trilogy of *Pre-Contract Practice* (first published in 1960), *Contract Administration* (first published in 1965) and *Tenders and*

Introduction

Contracts for Building (first published in 1975). All three books have been revised many times over the intervening years, the last such occasion being in 2003.

Much has changed since, so in this book not only have we brought together all of the material that appeared in those three publications, we have brought it up to date, we have edited out the inevitable duplications and we have added a chapter on value and risk management.

Despite these major revisions, we have aimed to continue the tradition of the Aqua Group and have tried to keep the concepts and terminology simple so the reader quickly grasps the essence of the subject matter. In this context, we have continued to use the terms *architect*, *quantity surveyor*, etc. without reference to gender and, to save repetition, we have reverted to 'he' where the sense requires. This, of course, is not to deny the existence of many able and eminent practitioners complying with the alternative specification! Indeed, we continue to echo the adage *vive la différence* – but we trust that our readers will understand the sentiment behind this decision.

All of the Aqua Group's books have assumed the use of the JCT Standard Building Contract. This book is no different and, except where noted otherwise, assumes the use of the JCT Standard Building Contract 2005 Edition. It is referred to throughout as the SBC – that follows the abbreviation given by the JCT itself – but most practitioners continue to refer to the contract as JCT.

... vive la différence ...

Success in building requires that a complicated series of interactions be completed in a logical and pre-determined sequence. Employers will then get the building they want, at the planned price, in the required time. In this book, we have aimed to set down the principles of good practice to be applied to the pre-contract and post-contract processes to make this goal more readily achievable.

We discuss those factors that need to be addressed by the building team in developing a project from the earliest involvement of a consultant through to the issue of the final certificate. We also discuss the effect of alternative procurement routes on the appointment and responsibilities of members of the design team. In the last three chapters we consider delays and disputes, insolvency and capital allowances.

We are grateful to the following organisations for permitting us to use their copyrighted material to reproduce some of their standard forms as our examples and to quote extracts from some of their published documents elsewhere in this book:

- **Building Cost Information Service Ltd** for their Standard Form of Cost Analysis elements.
- **Institute of Clerks of Works of Great Britain Incorporated** for their Clerk of Works Project Report form.
- **RIBA Enterprises Ltd** for their RIBA Outline Plan of Work 1998, Uniclass Section I, Architect's Instruction form, Interim Certificate and Direction form, Certificate of Practical Completion form, Certificate of Completion of Making Good Defects form, Final Certificate form and Notification of Revision to Completion Date form.
- **Royal Institution of Chartered Surveyors** for their Valuation form and Statement of Retention.

Given that we come from a quantity surveying background, it would not have been possible for us to update and edit this publication had we not had expert input from practitioners specialising in other areas of the construction industry. The practitioners concerned and the chapters which they contributed are as follows:

- Chapter 6: Value and Risk Management – Michael Dallas MA(Cantab), CEng, MICE, FIVM, MAPM, Partner of Davis Langdon
- Chapter 17: Drawings and Schedules – Peter Ullathorne JP, AADipl, RIBA, FRSA, FRSH, AAIA, Principal of The Ullathorne Consultancy
- Chapter 18: Specifications – Nick Schumann, Managing Partner of Davis Langdon Schumann Smith
- Chapter 25: Site Duties – Peter Ullathorne JP, AADipl, RIBA, FRSA, FRSH, AAIA, Principal of The Ullathorne Consultancy
- Chapter 32: Capital Allowances – Tony Llewellyn Dip QS, FRICS, Managing Partner of Davis Langdon Crosher & James

In respect of all of the above contributors, we count ourselves fortunate that highly experienced practitioners gave up spare time to let us have the benefit of their advice.

Once again, the drawings in the book are by Brian Bagnall. Sadly, we must record that Brian died in August 2004 so a rich source of wit and insight has been lost to the industry. We had the privilege of meeting Brian on a few occasions and marvelled at the speed of his pen when producing a cartoon to match any topic under discussion.

Finally, a debt of gratitude is owed both to the former members of the Aqua Group and to Madeleine Metcalfe, senior commissioning editor in Blackwell Publishing's Professional Division. Our thanks are due to the Aqua Group since they provided us with an excellent foundation on which to take this publication forward and to Madeleine whose management of the whole process made for a timely publication when, so often, we allowed the distractions of our work commitments to get in the way!

Mark Hackett MA, MSc, FRICS, ACIArb
Ian Robinson BSc(Hons), LLB(Hons), MRICS, ACIArb
Gary Statham BSc(Hons), MSc, MRICS, MCIArb

The Aqua Group
Oliver Burston
John Cavilla
Richard Oakes
Quentin Pickard
Geoffrey Poole
John Townsend
John Willcock
James Williams
and Brian Bagnall, now deceased

Part I

Briefing the Building Team

Chapter 1
The Building Team

Introduction

Over the years, since the first editions of the Aqua Group's books, the process of building has become increasingly complicated. From inception to completion, through site acquisition, design, contract and construction, each stage has become more time-consuming and thus more expensive. The need to optimise the process is of paramount importance and the best base from which to achieve this is proper and efficient team working. It is therefore vital that all members of the building team are fully conversant, not only with their own role but also with the roles of others and with the interrelationships at each stage of the project. Each can then play their part fully and effectively, contributing their particular expertise whenever required.

The make-up of any particular project team will depend on the scope and complexity of the project, the procurement route and the contractual arrangement selected. There are already many different methods of managing a project and, no doubt, others will be developed in the future. The following chapter is set in the context of traditional procurement and, although not exhaustive, will give an indication of the principles involved and the criteria by which other situations can be evaluated.

Parties to a building contract and their supporting teams

The parties to a building contract are the employer and the contractor. Those appointed by these two will complete the *building team* which can include:

The design team

- *employer
- project manager
- *planning supervisor
- *architect

- *quantity surveyor
- structural engineer
- building services engineers
- sub-contractors

In addition the employer may appoint:

- *clerk of works.

The construction team
- *contractor and/or principal contractor
- *site agent (or foreman, described in the contract as the person-in-charge)
- sub-contractors

It should be noted, however, that only those marked with an asterisk are mentioned in the Joint Contract Tribunal Standard Building Contract With Quantities 2005 Edition (SBC). This list is not exhaustive and to it could be added planners, landscape consultants, process engineers, programmers and the like. Also, some roles may be combined and roles such as the project manager or planning supervisor may be fulfilled by individuals or firms from varying technical backgrounds.

Rights, duties and responsibilities

The SBC is comprehensive on the subject of the rights, duties and responsibilities of the employer, the contractor and the other members of the building team mentioned in it. Not all the members of the building team are mentioned in the SBC and those not mentioned will usually be given responsibility by way of delegation from those who are mentioned. This delegation must be spelt out elsewhere in the contract documents, usually in the bills of quantities.

Whatever the size of the building team, each member should be familiar with the contract as a whole and, in particular, with those clauses directly concerning their own work, so that the project can be run smoothly and efficiently. It should be noted that some of the duties are discretionary, some are mandatory and some even have statutory backing.

The employer

The employer is referred to throughout the contract and is expressly required to perform specific duties. The vast majority of these duties are codified and are carried out by the architect/contract administrator on behalf of the employer. However, the employer, as one would expect, retains the important duty of payment to the contractor for works which are completed in accordance with the contract. Such is the importance of the payment provisions in the SBC, and indeed in all construction contracts, that failure to adhere to the provisions may lead to statutory repercussions against the employer.

The architect/contract administrator

The architect/contract administrator is named in the contract and, as the designation contract administrator suggests, is not only responsible for carrying out the design of the works but also the vast majority of the administrative duties under the contract on behalf of the employer. The architect/contract administrator is also the only channel of communication for any un-named consultants with delegated powers.

The quantity surveyor

Quantity surveyors are named in the contract and their principal duties are in relation to the payment provisions, value of the works including the value of variations and, if so instructed, ascertainment of the loss and/or expense suffered by the contractor as the consequence of a specified occurrence.

The status of named consultants

While the architect/contact administrator and the quantity surveyor are referred to in the contract and are expressly required to perform specific duties (many clauses include the phrase 'the architect/contract administrator shall'), they are not parties to the agreement. Should the contractor have a grievance, if they fail to carry out their duties as defined in the agreement, the only contractual recourse is to seek redress from the employer.

Unnamed consultants with delegated powers

The project manager and the structural or other consulting engineers are not referred to in the contract, nor do they have any express powers under the contract. They do, however, have a duty, as the employer's persons, not to impede the progress of the contractor. Their position within the building team depends on the agreement they have with the employer or the architect. Where they have been given responsibility by way of delegation, perhaps for design or site inspection, they should be named in the contract documents and the extent of their delegated responsibility should be identified so that they have contractual recognition. Since they have no powers

under the contract, if they need to issue instructions this must be done through the architect contract administrator.

As with the named consultants, if the contractor has any grievance against unnamed consultants the only contractual recourse is to seek redress from the employer.

The project manager

Project managers, who may be considered as employers' representatives, are likely to be appointed at the outset by the employer to whom they are directly accountable. The project manager is responsible for the programming, monitoring and management of the project in its broadest sense, from inception to completion, to ensure a satisfactory outcome. This will involve giving advice to the employer on all matters relating to the project, including the appointment of the architect, quantity surveyor and other consultants. However, not being mentioned in the contract, the project manager's position in respect of the contract works, which will be only a part of his overall duties and responsibilities, must be clearly determined and described in the contract documents. The duties and responsibilities delegated must not exceed those set out in the contract in respect of the employer.

The planning supervisor

The Construction (Design and Management) Regulations 1994 (CDM94), which came into effect on 31 March 1995, introduced the 'planning supervisor' to the building team. The planning supervisor is specifically referred to in the SBC as being appointed by the employer, pursuant to regulation 6(5) of the CDM Regulations and, under Article 5, the planning supervisor is stated to be the architect/contract administrator, unless some other individual is identified.

The planning supervisor is required to:

- ensure that the design avoids foreseeable risks
- ensure co-operation between designers
- be able to advise the client (as the employer is termed in CDM94) and contractor on the competence of appointed designers
- be able to advise the client on the competence of a contractor
- ensure preparation of a draft health and safety plan for the contractors at tender stage
- give notice of the project to the Health and Safety Executive
- ensure preparation of the health and safety file
- review and amend the health and safety file as work proceeds
- deliver the health and safety file to the client on completion of the project.

The principal contractor

In the SBC, under Article 6, the principal contractor is stated to be the contractor or such other contractor as the employer shall appoint as the principal contractor pursuant to regulation 6(5) of CDM94. The employer must be reasonably satisfied that such principal contractor is competent and can allocate sufficient resources to ensure compliance with the regulations. The definition of a principal contractor in CDM94 allows for considerable flexibility but for the majority of construction contracts carried out under the SBC the main contractor is appointed as the principal contractor.

Since there must always be a principal contractor in post while work is in progress on site problems could arise where enabling works contracts, such as for demolition or piling, are required before the commencement of the main contract, or where a fitting-out contract follows completion of the main contract. In these circumstances the employer will probably appoint a succession of principal contractors but care will need to be taken to ensure responsibility passes properly from one to the next.

The principal contractor is required to:

- ensure that the other contractors comply with their obligations on health and safety matters
- co-ordinate co-operation between sub-contractors
- ensure compliance with the health and safety plan
- ensure that all relevant persons comply with the health and safety plan
- ensure that only authorised persons are allowed on site
- display a copy of the notice submitted to the Health and Safety Executive

- provide the planning supervisor with the information required for the health and safety file
- instruct other contractors so that the principal contractor's duties are complied with.

Sub-contractors

Sub-contractors feature as members of the design team and as members of the construction team. This is because, as well as carrying out work on site, they are often involved in the design and planning of specialist works in advance of the appointment of the main contractor. Under the SBC sub-contractors do not acquire rights in the way that nominated sub-contractors previously did under that contract's predecessor, the Joint Contacts Tribunal Standard Form of Building Contract Private With Quantities 1998 Edition (JCT 98). All sub-contractors employed by the contractor under the SBC have 'domestic' status and are fully accountable to the contractor. The contractor has full liability for the sub-contract work.

Under the SBC's predecessor, the JCT 98 (which is still very much in use today), nominated sub-contractors acquire special rights to payment and have responsibilities which must be discharged in close liaison with the rest of the building team. In JCT 98, nominated sub-contractors' special rights to payment are safeguarded by imposing duties on the quantity surveyor, the architect/contract administrator and the contractor.

The clerk of works

The clerk of works is normally appointed by the employer to act, under the direction of the architect/contract administrator, solely as an inspector of the works. Traditionally, the role would have been taken by an experienced tradesman, perhaps a carpenter and joiner or bricklayer. However, with today's highly complex and high-tech buildings, the architect, who will normally recommend the appointment, may need someone technically experienced or qualified and here the Institute of Clerks of Works will be able to assist in finding the right person. The clerk of works should be ready to take up the duties before the date of possession (how early will depend on the size and complexity of the project) and will either be resident on site or will visit the site on a regular basis during the period of the works.

Statutory requirements

Under the SBC the parties and the various consultants are charged with many duties, some of which are discretionary ('the architect/contract administrator may . . .')

and some of which are mandatory ('the contractor shall . . .'). In addition, provision is made for compliance with duties and responsibilities imposed by legislation (statutory matters) as follows:

- clause 2.17 Divergences from Statutory Requirements
- clause 2.18 Emergency compliance with Statutory Requirements
- clause 2.21 Fees or charges legally demandable under any of the Statutory Requirements
- clause 3.25 CDM Regulations – undertakings to comply
- clause 3.26 CDM Regulations – appointment of successors
- clause 4.6 Value Added Tax
- clause 4.7 Construction Industry Scheme.

These may change from time to time and may be augmented by the addition of regulations issued by government under delegated legislation. Constant vigilance is therefore required to keep up to date.

The CDM Regulations

The Construction (Design and Management) Regulations 1994 (CDM94) place duties on clients, planning supervisors, designers and principal contractors to plan, co-ordinate and manage health and safety throughout all stages of a construction project. Since CDM94 came into force on 31 March 1995, the regulations have been amended a number of times by regulations such as The Management of Health and Safety at Work Regulations 1999, The Construction (Health, Safety and Welfare) Regulations 1996 and The Construction (Design and Management) (Amendment) Regulations 2000 (the latter amendments merely make minor amendment to the definition of designer). The Health and Safety Commission has frequently amended CDM94 due to its continued commitment to maintain the health and safety of construction workers and the general public.

It is worthy of note that contravention of the regulations can be a criminal offence. The regulations apply to most projects, but there are a few exceptions – notably, they do not apply to:

- construction work when the local authority is the enforcing authority for health and safety purposes
- work which is expected to last no more than 30 days or involve no more than 500 person days of construction work and involves less than five people on site at any one time – apart from Regulation 13 (designer duties)
- work for a domestic client, unless a developer is involved – apart from Regulation 7 (site notification) and Regulation 13 (designer duties).

However, where the construction phase of a project involves demolition, CDM94 applies regardless of whether or not the work would otherwise have been exempt.

Under the overall guidance of the planning supervisor it is the designer who is the responsible person in the early stages of a project and, while such duties will obtain throughout the design and construction periods, it is in the pre-contract stages that the majority of the designer's responsibilities under CDM94 are to be discharged. It is the designer's duty to tell the client (employer) initially what the client's duties are. The designer should design to reduce, if not avoid, as far as reasonably practicable, risks to health and safety, both during construction and during subsequent occupation, and ensure that the design documentation includes adequate information on health and safety. Such information should be included on drawings or in specifications and should be passed to the planning supervisor for inclusion in the health and safety file. The designer should not get involved with the principal contractor's duties – CDM94 is quite clear that the management of safety during construction is the responsibility of the principal contractor.

At the time of writing the Health and Safety Commission (HSC) has issued a consultation document reviewing CDM94. It is intended that the consolidated and revised Regulations are to be laid before Parliament in 2006.

Avoiding disputes

It is imperative that the building team implements formal procedures and proper documentation to succeed in the efficient and smooth running of building contracts – clarity and transparency are perhaps the key words. This is particularly relevant in the avoidance of disputes. As projects become more complex, costs increase, margins tighten and employers demand greater financial and quality control, the margin between success and failure narrows, and there is less flexibility to absorb the 'swings and roundabouts' which have ever been a feature of the construction industry. The likelihood of disputes arising is, therefore, increased.

The building team must be vigilant and ensure that the procedures employed, and the working relationships built up, produce an environment of co-operation rather than discord. Just as the building team strives for perfection in its working relationships, so the JCT strives to refine its contracts to take account of current practice and the latest decisions in the courts. Hence the frequent revisions to the JCT's standard forms of building contract which aim to achieve clarity and certainty in formalising the relationships between the parties.

In the event that disputes do arise, the importance of proper documentation and compliance with formal procedures cannot be over-stressed. If dispute proceedings become inevitable it should be some comfort to know that proper documentation will be an asset rather than a liability.

On the assumption that the contract documents are complete, that tenders are reasonable and reflect current prices, and that information is available when required and not subject to late changes on the part of the employer or the architect, the origin of contractual disputes is seldom found to be in the dishonesty or incompetence of any party but rather in the failure of one member of the team to convey information clearly to another. Unfortunately, what is clear in the mind of the architect, for example, might be 'misty' to the quantity surveyor and 'foggy' to the contractor. This can lead to all sorts of problems! Conveying intentions and instructions clearly is vital in the successful management of a contract and, all too often, it is the breakdown or failure of communications that is the root cause of a dispute.

Communications

Many books, conferences and papers have been devoted to the subject of 'communications' and it is a matter that cannot be dealt with exhaustively here. However, set out below are certain golden rules to be observed by all members of the team in their dealings with each other:

- Do not unnecessarily tamper with the standard clauses of the SBC, but if the client nevertheless requires it, employ a specialist.
- Ensure that the contract is executed prior to any start on site.
- Ensure that all team members have certified copies of the contract documents.
- Be realistic when completing the contract particulars and identify such information in the bills of quantities at tender stage.
- Issue all instructions to the contractor through the architect.
- Issue all instructions to sub-contractors or suppliers through the contractor.
- Use standard forms or formats for all routine matters such as instructions, site reports, minutes of meetings, valuations and certificates – preferably with sequential numbering.
- If instructions are issued other than in writing, ensure that they are confirmed in writing as soon as possible after the event.
- Ensure that everybody is kept informed, not just those who have to act.
- Be precise and unambiguous.
- Act promptly.

Examples of standard forms and suggested standard layouts for the more important communications passing between members of the building team are given in later chapters.

Chapter 2
Assessing the Needs

The structure

The successful development of any building project requires the acceptance by all parties of an underlying structure or framework of operation. For a traditional building project the Outline Plan of Work published by RIBA Publications and reproduced as Figure 2.1 is an excellent model framework for running a project. It also provides a useful checklist for key activities.

As work proceeds, it is prudent to reconsider earlier decisions to ensure that the developing design continues to satisfy the employer's overall criteria. Generally, the procedures recommended in this book follow the Outline Plan of Work. However, the plan can be modified to suit those procedures required by alternative methods of building procurement.

The brief

The initial brief from the employer will often be little more than a statement of intent; it may be no more than a telephone call. At this stage there is unlikely to be any formal appointment. After the initial euphoria surrounding 'getting a job', members of the design team will quickly realise that far more information is required from both the potential employer and from within their own organisations.

From the potential employer each member of the design team will, typically, require details of:

- whether or not they are in competition with others for the commission
- the other members of the design team
- the status of the employer – contractor, developer, purchaser or owner-occupier
- the reason for the employer's desire to build – accommodation requirement, business expansion, company image, or investment

RIBA Outline Plan of Work 1998

The Work Stages into which the process of designing building projects and administering building contracts may be divided. *(Some variations to the Work Stages apply for design and build procurement.)*

FEASIBILITY

A **Appraisal**
Identification of client's requirements and of possible constraints on development. Preparation of studies to enable the client to decide whether to proceed and to select the probable procurement method.

B **Strategic Briefing**
Preparation of Strategic Brief by or on behalf of the client confirming key requirements and constraints.
Identification of procedures, organisational structure and range of consultants and others to be engaged for the project.

PRE-CONSTRUCTION PERIOD

C **Outline Proposals**
Commence development of Strategic Brief into full Project Brief.
Preparation of Outline Proposals and estimate of cost.
Review of procurement route.

D **Detailed Proposals**
Complete development of the Project Brief.
Preparation of Detailed Proposals.
Application for full Development Control approval.

E **Final Proposals**
Preparation of Final Proposals for the project sufficient for co-ordination of all components and elements of the project.

F **Production Information**
F1 Preparation of production information in sufficient detail to enable a tender or tenders to be obtained.
Application for statutory approvals.
F2 Preparation of further production information required under the building contract.

G **Tender Documentation**
Preparation and collation of tender documentation in sufficient detail to enable a tender or tenders to be obtained for the construction of the project.

H **Tender Action**
Identification and evaluation of potential contractors and/or specialists for the construction of the project.
Obtaining and appraising tenders and submission of recommendations to the client.

CONSTRUCTION PERIOD

J **Mobilisation**
Letting the building contract, appointing the contractor.
Issuing of production information to the contractor.
Arranging site hand-over to the contractor.

K **Construction to Practical Completion**
Administration of the building contract up to and including practical completion.
Provision to the contractor of further Information as and when reasonably required.

L **After Practical Completion**
Administration of the building contract after practical completion.
Making final inspections and settling the final account.

Figure 2.1 RIBA Outline Plan of Work.

- whether or not the employer has a site and whether or not any preliminary discussions have taken place with the planning or other statutory authorities
- any preliminary work undertaken by the employer to establish a need or a market for the proposed project
- any technical skills that the employer can contribute
- any critical dates/timing.

From within their own organisations each member of the design team will need to assess:

- whether or not they are qualified to undertake the commission
- current workload – to help determine whether or not the commission can be serviced and whether new staff will have to be recruited
- the level of risk – there may well be an element of project development work for which no fee will be paid if the project does not proceed beyond a certain stage (for example, the design team will often be expected to undertake a degree of speculative work in assisting with the preparation of a site bid, particularly when the prospective employer is a developer)
- the effect of different procurement routes on their responsibilities and liabilities
- whether or not their professional indemnity insurance cover is adequate
- the employer's financial ability to complete the project.

The initial programme

One of the first questions the employer will ask will be: 'How quickly can I have my building?' At this very early stage it will be impossible for the design team to set a hard and fast programme. However, a simple bar chart (see Example 2.1) setting a timescale for the various key activities will provide both the employer and the design team with a useful framework within which to try to operate. The chart needs to identify separately any activities where the time-scale is beyond the control of the design team and the employer, for example in obtaining development control and other statutory approvals.

At this stage the employer has to rely on the professional judgment of his advisers as to the likely implications of any difficulties. The chart contained within this book has assumed a straightforward project which should not encounter major delays.

In setting the initial programme it should be recognised that it is unlikely that the procurement route (be it drawings and bills of quantities, design and build or management contract) will have been established. Therefore, any assumptions which have been made in the absence of firm decisions should be confirmed during the development of the brief and adjustments made to the programme accordingly. The assessment of the resources required for the project and the agreement to an initial programme will assist the design team in advising on the development brief and the preparation of the detailed programme.

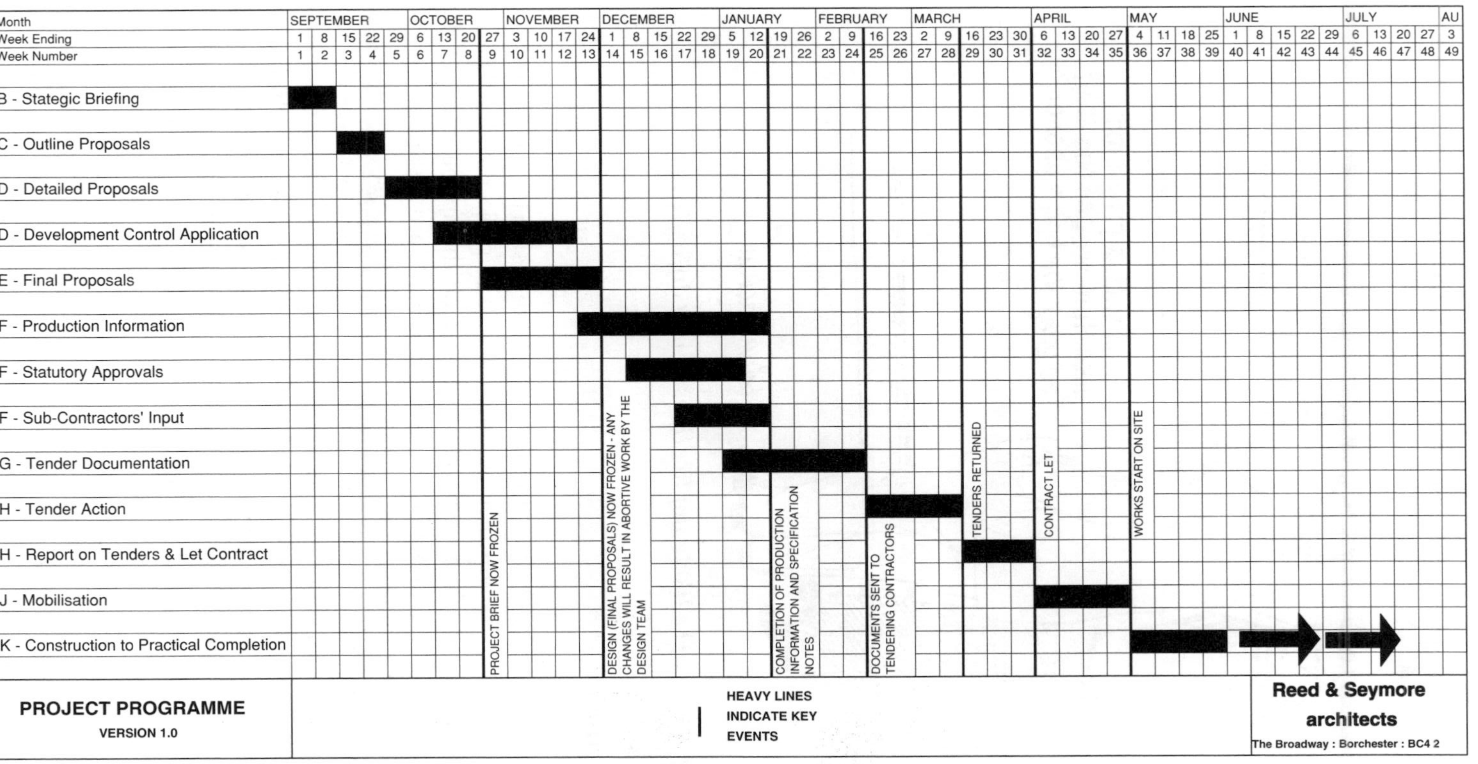

Example 2.1 The initial programme.

The likely implications of any difficulties

The appointment

Traditionally, an employer's first appointment to the design team is the architect contract administrator, who will then assume the role of project leader. The architect contract administrator will be expected to advise on the appointment of the other consultants to the team, although those appointments are invariably made direct with the employer. In this way the architect contract administrator avoids contractual liability for the professional performance of the other members of the design team and for the payment of their fees. Increasingly, this traditional arrangement is being modified to suit alternative methods of building procurement which may result in a different project leader, for example a project manager (see Chapters 1 and 15).

Most employers invite fee bids from consultants for the particular service required. However, even when the employer provides the fullest possible information, when few details of a scheme are certain the preparation of realistic fee bids at an early stage is both difficult and financially hazardous. Equally, the sensible evaluation of competing bids is difficult. Just as with the submission of an unrealistically low building tender, the employer should be cautious of the unrealistically low fee bid and question the consultant's ability to perform the required services satisfactorily – although in any competitive tender/fee bid situation, it is often the lowest price that is accepted. For those seeking to appoint consultants on the basis of quality rather than cost the Construction Industry Council (CIC) published in 1998 *A Guide to Quality Based Selection of Consultants: A Key to Design Quality*. The publication states that the CIC 'firmly believes that it is neither in the best interest of the client [employer] nor of the project itself that consultants be selected on the basis of a procurement system which includes a price comparison of their professional services.'

Appointment documents

Some professional institutions, such as the RICS or RIBA, make it compulsory for their members to provide written notification to their client of the terms and conditions of their appointment. Any member of the design team who belongs to an institution that does not so require is advised to confirm, in writing to their client, at least the following:

- the name of the client
- the services that are to be provided
- the conditions of appointment
- how fees are to be calculated and when they are to be paid
- terms that will apply should the project change or not proceed.

Most professional institutions publish guides for clients relating to the appointment of their members together with standard terms of appointment and forms of agreement for execution by the client and design team member. Typically these documents will also provide for the specification of the services to be provided, often by way of a standard tick list, and the fee to be paid for carrying out those services.

In May 2005 the British Property Federation (BPF) published a standard form of consultancy agreement for appointing professional consultants on construction projects. This standard form is one of very few forms to have been written from the perspective of the employer and, with some adjustments to each respective consultant's obligations, is designed to appoint all consultants, whatever their discipline.

Much of the criticism levied against standard forms of appointment is done so on the basis that each standard form is written to protect the liabilities of those seeking to impose it. Thus, the BPF form might be said to serve employers better than appointing an architect under a RIBA form.

Collateral warranties

In addition to the normal appointment documents consultants are frequently required to enter into collateral warranty agreements. Such agreements are used to create direct contractual relationships between parties other than the employer (e.g. funding institutions, purchasers, tenants) who are likely to suffer economic or consequential loss arising out of a construction project and any parties likely to cause such loss (e.g. consultants, contractors). The term collateral is used because the warranty is a side agreement to the warrantor's main contract with the employer.

The BPF and Construction Industry Council (CIC) both publish a series of standard Warranty Agreements. BPF standard Warranty Agreements comprise:

- CoWa/F (4th Edition 2005) – Collateral Warranty for funding institutions by a consultant
- CoWa/P&T (3rd Edition 2005) – Collateral Warranty for purchasers and tenants by a consultant.

The CIC standard warranty agreements comprise:

- CIC/ConsWa/F – Collateral Warranty for funding institutions by a consultant
- CIC/ConsWa/P&T – Collateral Warranty for purchasers and tenants by a consultant
- CIC/ConsWa/D&BE – Collateral Warranty for employer by a consultant working for a design and build contractor.

The need for collateral warranties stemmed from the English law doctrine of privity of contract which provides that a person who is not a party to a contract cannot enforce that contract even if the obligations under the contract are for their benefit. On 11 May 2000 the Contracts (Rights of Third Parties) Act 1999 came into force and reformed the law relating to privity of contract by providing a person who is not a party to a contract with a right to enforce a contractual term. A third party may enforce a contractual term if the contract expressly provides that they may do so, or if a term purports to confer a benefit on the third party by reference to membership of a class or answering a particular description, even if not in existence when the contract is entered into. Whilst the Act is now a part of English law, and will therefore apply to any contract governed by English law insofar as the contract expressly states or purports to confer a benefit, it can be excluded from any contract by the use of a simple contracting-out clause. Many of the standard forms, whether consultant appointments or construction contracts, do opt out of conferring third party rights. However, the JCT in the SBC affords users the opportunity to choose either third party rights or collateral warranties.

Whilst the Act, in theory, removes the need for collateral warranty agreements, this has not as yet happened in the construction industry – partly because of potential problems associated with the determination of the extent of a party's liabilities and the securing of appropriate insurance to cover them and partly because of unfamiliarity with the new law.

Part II
Available Procurement Methods

Chapter 3
Principles of Procurement

Simple theory – complex practice

Procurement, the process of obtaining goods and services from another for some consideration, is one of the oldest transactions known to man. It is in essence a simple transaction but today's construction projects are often vast in scope, involve several designers, require specialist contractors, are executed at arm's length and take considerable time to complete.

The complexity arises from the need to comply with numerous regulations, involve professional consultants, achieve value for money, demonstrate accountability, regulate complicated contractual relationships and through all this achieve a timescale largely dictated by the client's specific business objectives. Nevertheless irrespective of however complex modern construction projects may be, they are all based on the comparatively simple principles – cost, quality and time.

The eternal triangle

From a client's perspective, the COST, QUALITY and TIME paradigm might be considered as being the highest quality, at the lowest cost, in the shortest time. Unfortunately this is not always possible and a compromise has to be sought, based on the client's priorities. These criteria, and the compromise, can be visualised as a triangle (see Figure 3.1).

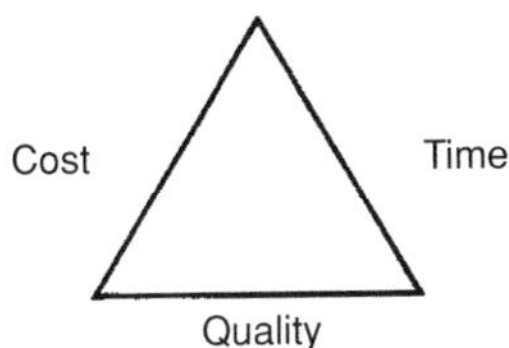

Figure 3.1 The procurement triangle.

Procurement triangle

If these three factors are kept in balance, with appropriate quality being achieved at an acceptable price in a reasonable timescale, the triangle would appear equilateral – with equal weight or emphasis being given to each factor. If, however, particular circumstances dictate that one of the factors must take precedence, the other two will 'suffer', or carry less weight or emphasis. The decision must lie with the client.

If quality is of paramount importance adequate time must be allowed for the design and specification to be perfected, and cost could rise on both counts. If speed of completion is paramount quality and cost may both have to suffer. If lowest cost is the priority, time may not be prejudiced but quality could suffer (see Figure 3.2).

The effect of these different priorities is relative, and there is no reason why, with proper planning and management, those elements with a lower priority cannot be adequately controlled.

It is advisable to apprise the employer of the procurement options, recommending the most appropriate route, detailing how the priorities are protected and outlining implications. Only when the method of procurement is determined can the

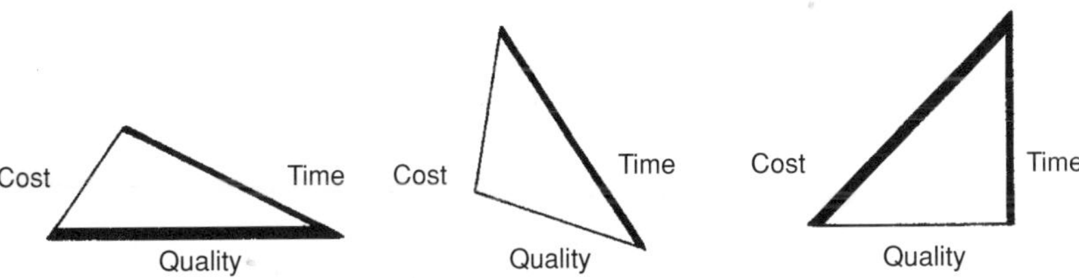

Figure 3.2 Variations on the eternal triangle, showing the different priorities.

pre-contract programme be finalised. This also represents the point in time when the specific services required of each member of the design team can be defined, fee proposals confirmed and agreements finalised.

One interpretation of the time, cost and quality priorities which may be derived from three different procurement routes is given in Example 3.1. It must be stressed that the particular circumstances of each project may result in different conclusions as to the procurement route that best satisfies the priorities of time, cost and quality. For example, a high level of provisional sums within a contract could undermine the cost certainty otherwise afforded by a particular procurement route. Additionally, it should be noted that, for ease of presentation, time allocation for building control approval has been omitted from Example 3.1. This element in itself can influence the choice of the procurement route.

A key decision when selecting a procurement strategy is based on the manner in which the detailed design is progressed. For example, by following a traditional procurement route the design team will develop the design, whereas with design and build procurement the design team may only prepare a design brief, the design itself being completed by the contractor. Each procurement option has a different time, cost and quality scenario (the procurement triangle).

The importance and weight given to each criterion is established in the briefing process and is a fundamental prerequisite to deciding which procurement route to embark upon. Even the best procurement strategies, once established having regard to a client's priorities, can be undermined by such things as:

- overly prescriptive briefs which give little flexibility to the design team
- unnecessarily high standards
- delays in decision making
- the use of one-off solutions rather than standardised products
- scope changes beyond those considered at the time of selecting the initial procurement route
- poor communication in the supply chain
- conflicting agendas and objectives
- doubling up of professional resources (as between the client and the contractor)
- fixed mindsets and recycling old solutions
- issue of design information which is late and/or inadequate and/or incomplete.

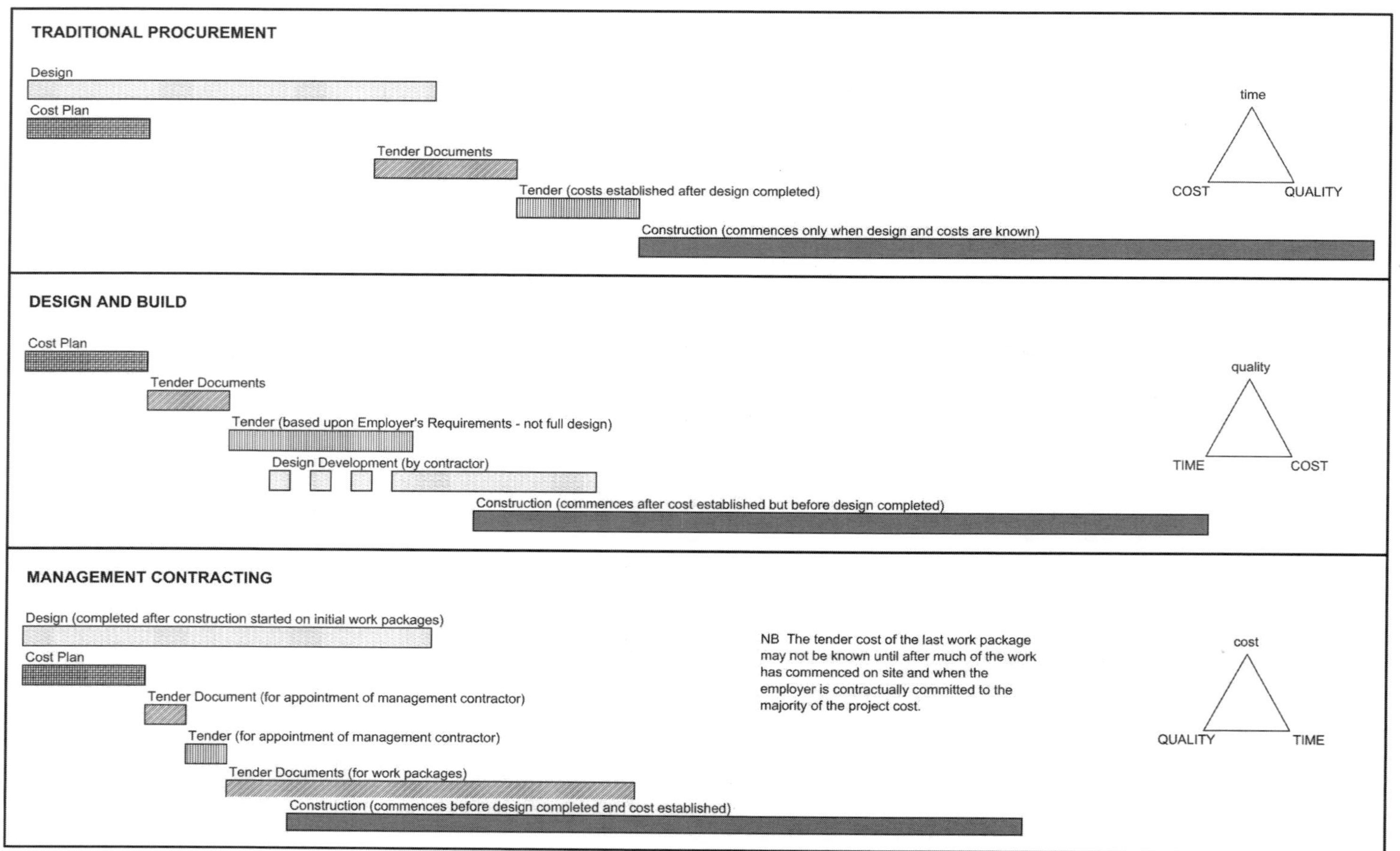

Example 3.1 Procurement options.

It is of paramount importance that the foregoing risks are recognised and to have, as part of the procurement strategy, management systems aimed at their eradication or, as a bare minimum, their control. Selection of a procurement strategy also influences the choice of which form of construction contract is used during the project. There are many standard forms available to complement any chosen strategy. It is worth noting, however, that the selection of any particular form of contract is a secondary consideration to selecting a procurement route – the selected form of contract merely provides the administrative framework through which the procurement process can be achieved.

Other considerations

Cost, quality and time are not the only criteria to be considered when selecting a procurement strategy. The design team should make the client aware of all the other factors which may influence the decision as to the most suitable procurement system, such as:

- The Construction (Design and Management) Regulations 1994 (CDM94)
- risk
- accountability.

The Construction (Design and Management) Regulations 1994

As Chapter 1 states, CDM94 places statutory duties on clients, planning supervisors, designers and contractors to plan, co-ordinate and manage health and safety throughout all stages of a construction project. However, the contractor is responsible for health and safety during the construction phase of the project. The extent of the roles and responsibilities of the contractor and employer respectively under CDM94 depends on the nature of each project, the make-up of the employer's organisation and the employer's ability to undertake certain obligations. It is this indirect influence that may affect the procurement strategy, rather than CDM94 dictating the procurement strategy itself. Items that indirectly affect a procurement strategy comprise:

- management of the project
- nature of the project and information available
- contractor selection procedure and his supply chain management
- activities with high risks to health and safety
- the employer's rules for site operations (e.g. the employer operating under a nuclear licensed site)

A key benefit of CDM94 is that the regulations, whilst mandatory, are sufficiently adaptable to suit whatever the procurement strategy.

Risk

Risk is defined and explained in detail in Chapter 6. At this stage, it is sufficient to note that decisions on the allocation of risk between the employer and the contractor can have a fundamental effect on the choice of the right tendering procedure and contractual arrangement.

It is, therefore, important to establish the client's attitude to risk at the earliest opportunity. If the client is prepared to accept increased risk, the consultants will have much greater flexibility to devise an effective strategy based on non-traditional procedures.

Accountability

Accountability is closely linked with risk and is considered in Chapter 5. However, decisions at an early stage can again have a significant effect on tendering procedures and contractual arrangements. It is therefore essential that the client identifies the level of accountability required on the project as early as possible, and for the consultants to explain the principles involved, so that an early decision can be made.

Making the contract

The process of making the contract may be divided into three parts:

- deciding on the type of contract and the particular terms and conditions under which the work will be carried out
- selecting the contractor
- establishing the contract price, or how the price will be arrived at.

Type of contract

The choice of the type of contract, and the particular terms and conditions under which the work will be carried out, will normally be made by the client in the light of advice received from professional advisers and will be inherently linked to the procurement strategy. The choice must be made at an early stage, as it will affect the way in which the contract documentation is prepared. In traditional single stage competitive tendering, the type of contract and the actual conditions to be used must be defined and the method of establishing the contract price must be decided before tenders are invited. In negotiated contracts, these decisions can be delayed until the contractor has been appointed.

The range of contracts available to choose from is considerable, and within each general type of contract a choice can be made as to the particular terms and conditions which are most suitable to the circumstances. There are two suites of contracts which cover most, if not all, domestic procurement strategies. These are the New Engineering Contract/Engineering and Construction Contract (NEC) family of contracts and the JCT 05 suite of contracts. By way of example, and identifying the vast array of main contracts available, the JCT suite comprises:

- Standard Building Contract (SBC)
  - With Quantities (SBC/Q)
  - With Approximate Quantities (SBC/AQ)
  - Without Quantities (SBC/XQ)
- Intermediate Building Contract (IC)
  - With Contractor's Design (ICD)
- Minor Works Contract (MW)
  - With Contractor's Design (MWD)
- Design and Build Contract (DB)
- Major Project Construction Contract (MP)
- Construction Management Agreement (CMA) and Construction Management Trade Contract (CM/TC)
- Management Building Contract (MC)
- Prime Cost Building Contract (PCC)
- Measured Term Contract (MTC)
- Framework Agreement (FA) and Framework Agreement Non Binding (FA/N).

To supplement the most widely used main contracts – namely, the SBC, IC, DB and MP – the JCT also produces supplementary sub-contracts for these forms, which give further flexibility to the chosen procurement strategy.

Selection of the contractor – the tendering procedure

The selection of the contractor and the establishment of the contract price, or how the price will be arrived at, may be combined in a single operation or they may be two separate operations.

First it is helpful to look at a definition of tendering that has held good for a number of years:

> *'The purpose of any tendering procedure is to select a suitable contractor, at a time appropriate to the circumstances, and to obtain from him at the proper time an acceptable tender or offer upon which a contract can be let.' (The Aqua Group)*

In fact, this is not so much a definition as a statement of the purpose of tendering. However, it needs amplification for a full understanding of all that it covers. The statement emphasises that the purpose of tendering is the selection of the contractor and the obtaining of the tender or offer.

Although any tender must take account of the conditions of contract under which the work will be performed, the tendering procedure is not dependent on the type of contract to be used. The contract conditions have the same significance as, for example, contract drawings and bills of quantities, which together make up the total contractual arrangements. It is important to appreciate this, otherwise considerable confusion will prevail. Contractual arrangements are concerned with the type of contract to be entered into and the obligations, rights and liabilities of the parties to the contract. These may vary because of the type of project, but there is no direct relationship between them and the tendering procedure.

For example, the same contractual arrangements could prevail in two cases where, in one, the contractor had been selected after submission of a competitive tender based on detailed drawings and bills of quantities, while in the other, he had been selected on the basis of a business relationship with the client and single tender negotiations. Chapter 21, Obtaining Tenders, contains a detailed explanation of tendering procedures.

Establishing price and time

There has, in the past, been an underlying assumption that the tendering process was only concerned with finding out which contractor could submit the lowest price based on the design presented to him. More recently, however, greater consideration has been given to situations where the contractor is partly or wholly involved in the design; where circumstances require that construction commences before the design has been completed; or where, for any other reason, price and time alone may not be the only factors to be taken into account.

In considering the purpose of tendering, one must think in the broadest terms. A tender quotes not just a price but also a standard of quality and a time within which the work will be completed. The purpose of tendering may not be for one job; it may be necessary to consider it in terms of a total programme, of which that job is just one project.

The dynamics of tendering

Tendering procedure today is a dynamic situation. It is not a procedure which can be applied across the board on all construction works; in fact 'procedure' is probably the wrong word, but, as it is used so frequently, it may be misleading not to use it here. There are so many different factors to consider in each case, and hence decisions to be made, that it would be more appropriate to refer to it as 'the tendering structure and requirement'.

Much of what has been written with regard to past attitudes to tendering procedures also applies to contractual arrangements. There has been great emphasis on fixed price contracts with a somewhat poor realisation as to what the word 'fixed' entails. Frequently the assumption has been that the contract should be a fixed price one, unless the situation is such as to make this impossible.

Although the actual form and the conditions of contract are part of the contractual arrangements, this book is not concerned with the legal contract as such. The choice of the actual form to be used on any given project is a matter to be decided in relation to the nature of the work and the needs of the client.

Finally, although everything written here is in terms of a contract between the client and the main contractor, it is equally applicable to a contract between a main contractor and a sub-contractor.

Chapter 4
Basic Concepts

Chapter 3 dealt with the principles of procurement and the aims and purpose of tendering and contractual arrangements. This chapter deals with some of the basic concepts to be considered in selecting a contractor and formulating arrangements whereby a satisfactory contract may be made, and looks at current trends in the industry. New legislation and new policies on government procurement, particularly in relation to funding arrangements, necessitate a constant review of existing practices and the development of new ones.

Sir Michael Latham's report *Constructing the Team* initiated a review of many of the traditional practices in the construction industry. It highlighted the problems arising from processes that are uncompetitive, inefficient and expensive, from relationships that are ineffective, and from methods of working that are adversarial. Add to this the problems of skill shortages, inadequate training and insufficient investment, and it is easy to understand why the industry has such a poor image. Whether the industry will achieve the 30% cost savings, targeted by Sir Michael, remains to be seen, but there is plenty of scope for improvement – it is hoped that the matters discussed in this book will help.

In Sir John Egan's report *Rethinking Construction*, resulting from a government construction taskforce, the cost reductions sought have been raised to 40%. While praising the industry for engineering ingenuity, design flair and flexibility, the report stresses the need for improvement through product development and by partnering throughout the supply chain. Partnering is referred to later in this chapter and considered in greater detail in Chapter 14.

Traditionally, time and money were the only considerations in selecting a contractor and making a contract with him. Whilst this is true in the broadest terms, there are many factors which affect either time or the financial outcome of the contract, but which are not, at first, apparent as being directly related to either.

Although not every project will offer opportunities for the introduction of new techniques in terms of procurement procedure, the design team ought to review the possibility. Matters worth consideration include:

- the economic use of resources – labour, materials, plant and capital
- the contractor's contribution to design and contract programme
- production cost savings
- continuity
- risk and accountability.

Economic use of resources

It may be thought that the money paid to the contractor is all that has to be considered – traditional single stage competitive tendering provides a good basis for monitoring this scenario. However, although money (the tendered price) provides a common basis of comparison of the extent of the resources to be used by the contractor, the client may still have an unsatisfactory situation. If the tendering procedure and contractual arrangements are badly drawn up, the contractor may be using his resources inefficiently. This is dealt with in greater detail later. To give a very simple example here: if a client has two similar projects, he will not necessarily get the lowest final price by setting up competition between one set of contractors for the first project and another set for the second and taking the lowest in each case. If the two projects are linked or combined in one, the price of the larger single contract may be less than the two let separately, because of the savings in building resources.

The first priority in tendering is to ensure the most economic use of the building resources, bearing in mind the particular needs of the client, and then to ensure that the price paid for those resources is as low as is reasonably possible.

The economical use of contractors' resources is becoming a much more difficult problem to assess today, as compared to when the building industry was mainly craft-based and tendering reflected little more than the contractors' addition for profit and their management skill in organising the output of their craftsmen to complete the job. In such a situation there was little opportunity for contractors to vary the deployment of their resources. With limited plant and machinery available and with production knowledge limited to the best way to organise gangs of craftsmen to carry out the work, it was possible for the contractor to be given the quantities and for the lowest tender then to represent the lowest production resources.

Today there is still much work that comes into this category but there are many elements in a modern building where different considerations apply. Even with the traditional elements of labour, materials, plant and capital, there are additional factors to be evaluated.

Labour

If labour is plentiful, a labour intensive form of construction will be economic, but, if labour becomes short, generally or in specific trades, difficulties may arise leading to inflated prices. It is also worth considering whether the emphasis can be shifted from site-based, to factory-based labour – a cold, wet and windy site in the middle of January is not the best environment to encourage good productivity from the workforce.

Materials

Not only should the use of new materials be considered, but also availability, bulk purchasing, special relationships between contractors and suppliers and special discounts. Life cycle costing studies have shown that increased investment in quality in original construction can reduce maintenance and replacement costs during the life of a building, leading to better economics overall.

Plant

A balance has to be struck between the use of larger factory made components, needing extensive cranage facilities with hard access, but with a consequent saving in site labour, and the use of small units, with minimum site plant requirements. Continuity, or the lack of it, in the use of expensive items of plant, has a significant effect on the economics of production.

Capital

Cash flow is vital to most contracting organisations – keeping loan charges to a minimum can significantly affect contractors' tenders. Keeping retention percentages to a minimum, and the prompt reimbursement of fluctuations in the prices of labour and materials are relevant here. However, a balance has to be struck between the maintenance of a reasonable retention fund and easing the contractor's cash flow.

There is, of course, another element that is vital in the quest for the economic use of resources. The most careful planning of the use and co-ordination of labour, material and plant in the design stages of a project will be of no avail if the contractor does not have the management expertise to bring the master plan to reality in practical terms. This is particularly important in the planning and co-ordination of specialist subcontractors' work. It is this aspect of project planning that has led to the extensive use of pre-tender interviews with prospective contractors – and even to partnering, which will be referred to in greater detail below.

Contractor's contribution to design and contract programme

There are occasions today when the contractor has a contribution to make which affects the design, speeds up construction, uses fewer production resources and, therefore, produces a more economical job. The importance of this contribution should not be overstated, as a substantial majority of building works are, and will continue to be, small in value, using fairly traditional methods. However, it is a factor not to be ignored and it is interesting that, where sub-contractors are concerned, there is almost an undue readiness to consider the sub-contractor's contribution, whilst that of the main contractor is often accepted with reluctance. This dichotomy is partly due to the fact that sub-contractors tend to represent specialist activities entailing considerable design responsibility, whilst main contractors frequently show reluctance to accept any such responsibility.

Engineering works differ somewhat from building works, in that the responsibility for detailed design is more frequently left in the hands of the contractor.

Cost planning techniques assist architects in designing economically. However, it is not always possible for the quantity surveyor or other members of the design team to know the best way in which the contractor's resources can be used to improve productivity in terms of cost and speed, particularly where a proprietary system is involved. For example, where a complete structure is fully designed before the contractor is chosen, the use of a proprietary system might well be excluded, despite its obvious economy, because of the subsequent amendments that might have to be made to the design in terms of detailed fixings and finishes. It therefore becomes necessary to consider, at the pre-tender stage, whether a contractor has a contribution to make or not, and if he has when this should be introduced.

Production cost savings

The economics of the construction industry's production methods are clearly a most relevant factor in the cost of building and obtaining value for money for the industry's clients. In terms of a particular project, this comes down to the economic use of the production resources held by a particular contractor. Production resources are dynamic, not static; production savings are continuously being made in the economy and the building industry is no exception.

Most manufacturing organisations have significant incentives to make production cost savings as it is the organisations themselves that benefit. In the construction industry methods of production are frequently limited by design and, therefore, cost savings are confined to more productive methods of constructing a given design. It is difficult but not impossible to promote cost savings where a design has to be altered to achieve them. Legal liability for a design remains with the designer and this might go some way to explain why there is a natural reluctance to make any such amendments which could increase the designer's risk. There is no reason why, however, a properly constituted value management exercise, undertaken during appropriate stages of design or construction, will not reveal some sort of cost saving. Ordinarily, unless special contractual arrangements are devised, a contractor will not have the incentive to look for productivity savings that affect the design.

'... *incentive* ...'

Where the opportunity of recognising production cost savings is considered to be an essential ingredient of a procurement programme special arrangements have to be made, including creating the incentive for the contractor to do so. This is often done in conjunction with continuity contracts which are discussed in Chapter 13. Although this may seem to be a daunting task, it is only an extension of a contractor's normal bonus system so contractors are very familiar with the principles.

Continuity

Continuity of activity is perhaps one of the most important ways in which production and management resources can be used economically and, provided the tendering procedures are right, the client can benefit. The building industry has a unique problem in relation to the provision of continuity in that each building is on a different site, and frequently user and client requirements will vary as well. Nevertheless, a great many buildings are for clients who will want similar facilities on similar sites and, furthermore, at least some of the detail in dissimilar buildings will be the same.

Everyone takes longer to carry out a task the first time. Big savings arise the second time, rather smaller ones the third and so on until, after carrying out the activity many times, virtually no further savings can be achieved. A succession of one-off situations, in management and production, is bound to be uneconomical, and this is one of the biggest single areas in construction where better use of production resources can be obtained with careful planning.

This factor has a very substantial effect on tendering procedures for, if continuity from a particular contractor is to be achieved, then it must be on a basis other than normal competition. This may make it necessary to plan the contractual arrangements for an initial project with a view to further projects being carried out by that contractor. Obviously, productivity savings arising from such continuity must be translated into a corresponding benefit to the client. Continuation contracts, serial contracting and term contracts are methods of doing this and are dealt with in Chapter 13.

Risk and accountability

Risk is also an important factor to consider in the tendering procedure and contractual arrangements. It is sometimes thought that, provided the contractor takes all the risk, accountability is satisfied. Frequently, however, this satisfaction is delusive, the true situation being hidden.

Risk is defined as the possible loss, which has to be stood by someone, resulting from the difference between what was anticipated and what finally happened. It is important to realise that a building owner takes on a considerable risk merely by the act of commissioning a building at all. The contractor never takes the total risk. The client, for example, takes the risk in relation to such matters as what the authorities will permit him to build on the site. The design team take the risk for the work they do, and so on. The question to be decided in the tendering procedure and contractual arrangements is how much risk it is appropriate for the contractor to take.

It does not always pay the building owner to ask the contractor to take all of the risk. It might be more beneficial that an insurance company, whose very purpose is to take the risk, does so and thereby charges a premium for doing so. There is no doubt that in every contractor's tender there is a hidden premium charging the client for the risk he has been asked to take, and in some cases, where the cost of the risk is very high but the probability low, it will not be worthwhile for the client to pass the risk to the contractor, just as in some cases it is not worthwhile insuring against the risk.

It is significant that government does not pay premiums to insurance companies to take the risk of destruction of government buildings by fire. This is good practice in terms of accountability for public money. Such are the resources of government, there is no need to pay others to take the risk. On the other hand, for small organisations, the destruction of a substantial asset can be disastrous and they must pay someone else to cover the risk for them.

In building, risk is much affected by the state of the client's own resources. It would be wrong to push the analogy with insurance too far. It is often difficult to distinguish the risks that is appropriate for contractors to take, as an inherent part of their production activities, from the risks inherent in the construction itself. It is, of course, entirely appropriate that contractors should take the risk for activities which are entirely under their control. It is appropriate, too, for the contractor to take the risk in operations, which are basically under the control of the building owner or his designers, where the building owner, because of his own circumstances, wishes the contractor to include in his tender a premium for taking that risk.

To illustrate this point, take on the one hand the conversion of an existing building where considerable risk is involved owing to the nature of the work. If the risk is too great, it may be impossible to get the contractor to tender at all. If it is moderate, then it would probably pay a client with substantial financial resources to take the risk himself and let the contract on a cost reimbursement basis. On the other hand, an individual of limited means, carrying out a conversion of his own house, may feel it is essential to get the contractor to take the risk and therefore get a fixed price.

A very common risk area today is that of fluctuations in prices of labour and materials, with the question as to which party should bear the risk involved.

Risk is one of the fundamental factors to consider in selecting suitable tendering procedures and contractual arrangements. It is a factor which depends on the client's circumstances as well as on the nature of the risk.

Accountability

An early reference to accountability was made in the above section considering risk, and the two subjects are very closely linked. However, accountability also has much wider implications and a full discussion is set out in Chapter 5.

Summary

For convenience, set out below are the five fundamental factors to be considered in the selection of a contractor and the type of contract to be used. To some extent they overlap and they certainly interrelate.

- the economic use of building resources
- the assessment of the contractor's contribution in relation to the design and speed of construction
- the incentives to make production cost savings and their control
- continuity of work in all aspects
- risk and the assessment of who should take it.

In addition it is worth noting the distinction between the selection of the contractor and the determination of the contract price. The traditional way of selecting a contractor, that of full drawings coupled with a bill of quantities describing and quantifying the work to be done, results in the selection of the contractor simultaneously with his price being known. In this situation, the drawings and bills of quantities are tendering documents and also rank as contract documents. Selection and determination of price occur together.

In many other situations this is not possible. What in fact happens is that the selection of the contractor is made, but only the basis on which the contract price is to be obtained is agreed. The final determination of that price may take place many months later. Indeed, it does not have to be determined before the contract is made. If this point is appreciated, some of the more unusual methods of tendering will become easier to understand.

Finally, it should be emphasised that although all these factors should be considered for every project or group of projects, it is recognised that in a great many small projects, particularly those which are completely one-off and not part of a larger general programme, some of them will hardly apply. Tendering procedures must be considered in the context of the national scene. Most building contractors are relatively small and the majority of builders would be unable and unwilling to tackle a job of any size. Many small contractors, while making a first class job of constructing a building to an architect's design, have neither the wish nor the ability to make the kind of contribution that national or specialist firms might make.

Chapter 5
Accountability

Accountability is a further factor which can have a profound effect on the choice of procurement method and such is its special nature that it deserves a chapter on its own.

Accountability arises whenever one party (the agent) carries out an activity on behalf of another (the principal or employer). The agent must account to the principal for the actions he takes. The extent of that account (or level of accountability) depends upon the principal's original brief and the degree of authority and responsibility delegated to the agent. In the procurement of buildings that authority and responsibility lies within a framework of established principles, with appropriate priority being given to cost, quality and time.

Background

Historically the emphasis has been on accountability within the public sector, although it is just as relevant in the private sector. The difference is purely one of reporting lines. In the private sector the relationship between agent and employer is direct: any matters can be settled between the two parties. In the public sector the responsibility of the agent to the employer (the public) is through an intermediary, usually an elected body, who will normally be advised by professional staff and who will ultimately be monitored by an independent audit body.

The accountability of the professional adviser is an important aspect of the tendering procedure in construction, because it is seldom possible for building owners to see what they are buying before it is built. Even when this may be possible to some extent, for example in the case of a standard building, each project is prototypical in nature due to the locality of its site.

Accountability differs from the other factors concerned with the tendering process, all of which are technical aspects, requiring professional advice to determine their effect. Accountability is a matter outside the strict constuction process since it applies wherever an employer's agent is acting on the principal's behalf in respect of purchasing, e.g. buildings, vehicles or stationery, etc. Clearly, therefore, it is a matter on

'... a matter on which the employer should have an opinion ...'

which employers should have an opinion and which should be considered to be part of their brief. Professional advisers, as agents, are responsible and accountable to the employer for both their actions and performance.

The modern concept of public accountability

The traditional concept of public accountability dates back to the Middle Ages. At that time production was craft orientated and the sophisticated methods of modern production and construction were unknown. Problems related to production were insignificant. If the State voted to spend money on ships for the navy, the problem uppermost in mind was to ensure that the money voted for the ships was actually paid to the shipbuilder, rather than finding its way into somebody else's pocket. The method used in those days for the placing of orders was probably simple bargaining, implicit in which was the objective of value for money. The simplicity of the whole operation served to highlight the only points where the matter could go seriously wrong; these related entirely to graft and corruption.

As methods of construction have developed it has become more important to be able to show that value for money has been obtained. While the standard of honesty of public officials is generally high, the constraints imposed upon them to ensure that there is no corruption tend to work against the concept of value for money. Indeed, it is possible for all the strictest canons of public accountability to be adhered to and yet extremely poor value for money to be obtained due to the waste of resources inherent in present public accountability procedures.

The main reason for this is that, with complex technological production methods, the inefficient use of resources may easily arise if not fervently controlled. To avoid such waste positive steps must be taken to ensure the optimum use of contemporary procurement and production techniques. These steps may be taken in value management/risk management workshops as covered in Chapter 6.

Contract documentation

The amounts involved in construction work are usually large and this means that money can be more easily misappropriated – thus there is more opportunity for corruption than in many other areas of activity. Contract documentation for construction work is therefore very important – it is vital to the concept of obtaining value for money and may be complex and technical, varying with the system of procurement adopted (see also 'Dispensing with competition' below).

Proper price

Clients should have an assurance that they are getting value for money or paying the proper price for a project even though it may not be possible to determine its total extent, or final price, at the time of the signing of the contract.

Where tenders are sought on a competitive basis, with full contract documentation, the lowest price is normally accepted as being the proper price for the work since all tenders are related to a common base. Where full documentation is not provided (e.g. in plan and specification or design and build contracts), although the basis of tendering is still common, it can be more difficult to convince a client that the proper price is being paid, because the precise specification may not be clear at tender stage.

Dispensing with competition

There are times when the use of competitive tendering is not possible, or will not necessarily produce value for money. There may be a number of reasons why the contractor should be selected without competition, and they may be complex and intangible. In many cases, it may be difficult to prove that there will be a monetary saving without the use of competitive tendering and the agent will have to convince the principal by justifying his professional judgment.

Where a contractor is selected on a basis other than equal competition, the contract sum or pricing mechanism must be negotiated and the matter becomes even more complex, especially if full documentation is not available. Not only does the selection of the contractor have to be justified, the concept of the proper price, or value for money, has also to be demonstrated.

Here again the documentation is vital – depending on the time and information available, a system must be set up that will not only allow a contract sum to be calculated but will also establish a pricing structure for the assessment of interim payments and variations where appropriate.

By their very nature, negotiated contracts often involve uncertainty, and the agent has a special responsibility to the principal to explain or justify the professional judgment that has led to the method of procurement recommended. In the public sector, whether in central government, local government or other government controlled organisations, it can be difficult to make a convincing case when accounting for the decisions thus made.

Inflation

A further factor related to accountability arises from inflation. Particularly in times of high inflation the argument has sometimes been advanced that the tendering procedure should be arranged so as to bring forward the commencement of building in order that an earlier price level may be obtained (to beat inflation). This may be a real saving to a private client depending on his circumstances and the use to which his money is put in the meantime. To government, however, whose concern is the percentage of resources devoted to building there is no saving at all, as inflation affects both sides of the balance sheet. However, this fact needs to be weighed against other factors obtaining at the time; public finance accounting procedures and the way in which central government organises its annual financial allocations may well place a premium on early financial year starts within the public sector.

Value for money

Value for money, as a concept, may be defined as achieving the optimum use of resources – resources comprising money, manpower, time and materials, each of which may be regarded as equally important. Until comparatively recently money was often regarded as the most important resource in the public sector, the cheapest method of construction being the one most usually selected. With money becoming more 'expensive' the need to obtain more for a given amount became imperative, and this has tended to emphasise the equal importance of constituent resources. For example, if a contractor is overpaid £1000 on a contract through an accounting error, there is a monetary loss of £1000; if, through design errors which do not manifest themselves during the contract, there is a loss of £1000 of building labour and materials then again there is a monetary loss of £1000. However, in the first case the money represents only financial resources which are no longer available for beneficial use, but in the second case the money represents not only a financial loss but also a loss of physical resources which have been used, wasted and lost.

Within both the public and the private sectors, the overpayment loss will be obvious; the loss of manpower and material resources may not. In either case, the professional advisers will be accountable to the building owner, to whom they are responsible, but whereas the former can be expected to be easily identified and recovered, the latter is considerably more difficult to detect and may never be recovered.

Government is concerned with the economic use of resources by the country as a whole. More resources used in building each project means fewer projects. While government is concerned with the monetary waste inherent in overpayment, this is very much the lesser of the two evils. Unfortunately, traditional systems only bring the overpayment to the fore, while the loss of resources remains obscure. Politically, too, overpayment is an easy matter to grasp and is highlighted in political debates as an example of government ineptitude, while the loss of resources is difficult to identify and for this reason is more often forgotten or ignored.

The definition of value for money has broadened over recent years. As money has become progressively dearer, so the concept has widened beyond that of obtaining best use of resources on site (usually by acceptance of the lowest tender) to embrace whole life cost of a project from initial procurement, through design and construction, to maintenance and cost in use and, perhaps, the cost of decommissioning in order that the best use/returns can accrue over the lifetime of the building. With this change in definition, terms such as 'life cycle costing' or 'whole life cost' and 'value engineering' have come into vogue, with value defined as 'the optimum balance of time, quality and cost'.

Summary

The problem, particularly with public building, is to identify whether or not there will be a saving in resources in relation to the way in which the contractor is selected. This is a complex and difficult task but in the long run the advantages to be obtained from many of the situations referred to later in this book will not be forthcoming unless the attempt is made.

Accountability is an extremely important matter in tendering procedures and contractual arrangements. It is essential that it is looked at in terms of all the resources used and not simply in terms of the price paid. The client's professional advisers must identify the effect that the procedure for the selection of the contractor may have on the use of all building resources and ignore, or certainly question, the historic traditions relating to public accountability.

Chapter 6
Value and Risk Management

Construction projects are about more than simply delivering buildings. They must reflect the long-term business needs of those who commission them and deliver the expected benefits.

In this context 'value' lies in the effectiveness with which the benefits are delivered. Effective delivery requires that the supply side of the industry clearly understands clients' long-term needs and delivers them efficiently and economically. Many initiatives focus on optimising the efficiency of delivery through the supply chain. 'Value management' complements efficient delivery by ensuring that efforts are deployed towards delivering the right buildings. It helps create affordable buildings that are productive and pleasant places to work in, that are easy to use and maintain and that contribute positively to the environment and communities in which they are situated. By contrast, effective 'risk management' ensures that value is not eroded by avoidable mishaps or uncertainties. To achieve this aim the team must, from the outset, ensure that conditions are put in place to enable successful project delivery.

This chapter describes how value management provides the means to articulate and deliver best value, whilst risk management provides assurance that the set-up and delivery of the project will avoid the destruction of value. It argues that both should be integrated with other project management processes to maximise the likelihood of a successful outcome.

Value management

Value management comprises a systematic process to define what value means for clients and end users of a facility, to communicate it clearly to the project delivery team and to help them maximise the delivery of benefits whilst minimising the use of resources. One of the most significant techniques used within the value management process is 'value engineering'. This is a structured technique that is applied to a design to deliver the required functionality at lowest cost and hence provide best value for money.

At the outset of a project, value management provides an exceptionally powerful way of exploring clients' needs in depth, addressing inconsistencies and expressing these in a language that all parties, whether technically informed or new to the construction industry, can understand.

This results in the following benefits:

- It defines what the owners and end users mean by value and provides the basis for making decisions, throughout the project, on the basis of value. It provides a means for optimising the balance between differing stakeholders' needs.
- It establishes the value profiles as the basis for clear briefs which reflect the client's priorities and expectations expressed in a language that all can understand. This improves communications between all stakeholders so that each can understand and respect the constraints and requirements of the others.
- It ensures that the project is the most cost-effective way of delivering the business benefits and provides a basis for refining the business case. It addresses both the monetary and non-monetary benefits.
- It supports good design through improved communications, mutual learning and enhanced team working, leading to better technical solutions with enhanced performance and quality where it matters. The methods encourage challenging the status quo and developing innovative design solutions.
- It provides a way of measuring value, taking into account non-monetary benefits, and demonstrating that value for money has been achieved.

Table 6.1 identifies three distinct phases in the application of value management to a construction project.

Value articulation and project definition

The use of functions to describe the benefits expected of a project enable the team to build direct links between the project objectives and the design solutions that later get built. These functions are often referred to as 'value drivers', since they

Table 6.1 Three phases in applying value management.

Project stage	Focus of activity
Inception to feasibility	Value articulation and project definition
Design and construction	Optimisation of benefits and costs
Commissioning and use	Learning lessons and performance optimisation

encapsulate what creates value for the client and end users. Essentially, building a function model of the project helps the members of the project team do three things:

- First, they can agree a clear description of what the project must achieve in terms of the benefits expected by the client and the end users.
- Second, they break this down into simple functional statements that describe the levels and quality of those benefits.
- Third, they break these down into clear statements that communicate to the design team those things that must be taken into account when the designs are developed.

Taken together, these statements describe the value that is expected at the outcome of the project and inform the development of the brief. They also provide the criteria upon which decisions should be based and feasibility stage options selected.

Optimisation of benefits and costs

Using the function model described above, the design team can develop designs that accurately reflect what the client and end users expect. However, satisfying the benefits in full represents only part of the optimisation of value.

To deliver good value for money the team must make best use of the resources that are available. This is where techniques such as value engineering are useful. This technique builds upon the function model developed in the project definition stage to the point where it identifies the proposed building elements that are linked with each function. By adding the estimated costs of providing the elements that contribute to each function the team builds up a picture of how much each function costs. By comparing this with the importance of each function, team members can assess where they are getting good value for money and where they are not.

The team generates ideas for performing the functions in different ways, starting with those that appear to offer lower value for money. Team members evaluate the relative merits of the ideas and develop those with most promise into detailed proposals for improvement. Finally, they submit their recommendations, based upon the proposals that they have developed, to the decision makers who will decide which to include. The technique is quick and very effective, provided the team applies it rigorously and is not tempted to take short cuts.

Learning lessons and performance optimisation

Once the construction stage of the project is completed the team should conduct a 'project review'. This provides the opportunity to check with the team and the building's users whether the full benefits that were defined at the outset of the project and

Table 6.2 The stages of value management.

No.	Type	Typical question answered
VM 0	Need verification	Is this the right project?
VM 1	Project definition	What are the project objectives?
VM 2	Brief development	What is the best option?
VM 3	Value engineering	Is this the most cost-effective solution?
VM 4	Handover review	Did we achieve our expectations?
VM 5	Post-occupancy / use review	Is the business sustainable?

Table 6.3 Study structure.

Stage	Study objectives
Preparation	Gathering and analysing information
Workshop	Workshop briefing, function analysis (and option selection, if appropriate) Creativity – generating ideas for improvement Evaluation – selection of the best ideas for development into proposals Development of proposals and action plans
Reporting and Decision Making	Presentation and reporting to obtain decisions on implementation
Implementation	Implementing the actions and proposals

predicted in the value engineering proposals were realised. The review can explore what went well and what could have been improved and provide the opportunity for particularly successful generic value management proposals to be 'banked' for use on future projects.

Table 6.2 identifies how the foregoing stages utilise the following value management study types. (See also Figure 6.2 for how these studies fit into the integrated process.)

The format for each study will usually follow the stages of a typical Value Engineering study, tailored to suit the project stage and the study objectives. Each study comprises four stages, identified in Table 6.3.

Risk management

Risk management is a systematic process to identify, assess and manage risks in order to enhance the chances of a successful project outcome. Risk management can also identify opportunities to enhance value. Like value management, the focus of risk management studies evolves with the project. Table 6.4 illustrates this evolution.

Table 6.4 Evolution of risk management studies.

Project Stage	Focus of the Risk Management Study
Inception	Are the risks acceptable?
Strategy	Are conditions in place to proceed?
Feasibility	Are risks allocated appropriately?
Pre-construction	Are risks under control?
Use	What can we learn for the future?

Risk must be managed

Articulating and optimising value throughout the design development stages of a project is of little benefit if risks run out of control and manifest themselves to undermine the successful delivery of value. It is therefore necessary to set up a process by which risks may be managed effectively.

In his report, *Constructing the Team*, proposing improvements to the construction industry, Sir Michael Latham stated 'No construction project is risk free. Risk can be managed, minimised, shared, transferred or accepted. It cannot be ignored'. Indeed, it is necessary to take risk if one is to maximise the benefits (or value) in an organisation. The first and major benefit of risk management, therefore, is that it enables senior management to embark upon projects in the full knowledge that they will be able to control risk and thereby maximise rewards. Engaging in fire fighting, whilst it may be exciting, is not an efficient way to control risk. It concentrates the attention on day-to-day matters whilst diverting attention from the wider issues. Risk management helps the team to concentrate on the big issues and manage these in an orderly way.

A formal risk management process delivers the following benefits for the project team:

- It requires that the management infrastructure is in place to deliver successful outcomes. This includes setting clear, realistic and achievable project objectives from the outset.
- It establishes the risk profile of the project, enabling appropriate allocation of risk, so that the party best placed to manage it has the responsibility for doing so. Risk allocation is a key component of contract documentation.
- It allows the team to manage risk effectively and concentrate resources on the things that really matter, resulting in risk reduction as the project proceeds. It also enables them to capitalise on opportunities revealed through use of the process.

Nothing ventured nothing gained

The saying 'nothing ventured, nothing gained' neatly captures the principle that taking risks is necessary in order to gain rewards. Thus project risk management is not about eliminating risk altogether but controlling the risks to which the organisation is exposed when undertaking a project from which defined benefits are expected. In order to control anything we need to understand it.

The first job in risk management, therefore, is to understand the project in depth. The following paragraphs explain the underlying concepts of risk management by going through the essential steps of the process.

Understanding the project

Listed below are some of the key questions that should be asked before embarking on any risk management study. In fact these questions are the same as should be asked when embarking on a value management study. This is unsurprising since both must begin with a thorough understanding of the project.

- Why is the project being undertaken?
- What are its objectives?
- What are the benefits that are expected to flow from it?
- Who is involved?
- What are their interests?
- Within what business environment is it being undertaken?
- What are the financial parameters relating to this project?
- What is the timescale?
- What are the constraints?
- What are the factors that are critical to its success?
- What design solutions are being proposed?
- How much it is expected to cost?
- Is there enough money available to pay for it?

Risks

Having gained answers to these questions one is well on one's way towards a comprehensive understanding of the project. One can then begin to identify what could prevent the full realisation of benefits. These are the risks to the project.

Opportunities

At the same time as identifying what creates uncertainty in the delivery of the benefits, one needs to understand the assumptions that have been made. These will normally have been embedded in the project business plan. Some of these could be overly cautious and, if the risk were not to materialise to the extent first anticipated, the outcome will actually be better than expected. These represent opportunities to increase project value and can be identified through risk management. Having identified such opportunities, it is wise to associate them with value management activities and thereby ensure that both techniques contribute towards increasing the project value.

Consequences

Having identified the risks, one needs to understand what would happen should those risks occur. These are the consequences of a risk occurring and are the things that could cause damage. The risk itself can remain a threat throughout the project but never cause any damage (apart from sleepless nights!).

Impact

It is only when the risk occurs that it will have an impact on the project. The impact may be financial, it may be related to the duration of an event, or it may affect the quality of the product. It may, of course, affect all three. For each risk it is necessary to assess, qualitatively or quantitatively, the impact of its consequence on each of these dimensions.

Likelihood

In order to understand how any risk might be managed, one needs first to understand how likely it is that it will occur. Again, this can be expressed qualitatively or quantitatively – qualitatively as 'likelihood', in the absence of actuarial data, and quantitatively as a 'probability' when actuarial data is available. Where something that may affect the project outcome is certain to happen, it is referred to as an 'Issue'. Issues need to be addressed by the project team in a similar way to risks.

Severity

The product of the impact and the likelihood of a risk provides a qualitative measure of the severity of the risk. This is generally sufficient to identify which risks require active management.

Risk management strategies

There are four major strategies for dealing with risk, and these are referred to in literature under various acronyms. Some use the acronym ERIC, described below, and a commonly used alternative acronym is TTTT, also shown below in italics.

- Eliminate (*Terminate*) show-stoppers and the biggest risks. This action will be taken only with the most severe risks which threaten to cause project cancellation or failure. For these risks it is essential that the risk be removed entirely or reduced to a less severe risk to enable one of the other strategies to apply. There are several ways to do this. The team could review the project objectives and remove or re-think the objective which contains the intolerable risk: the result will change the project. The team could reappraise the whole concept of the project, again creating a changed project. It might be decided that the risk is so great that the project is not a viable proposition: this would result in cancellation.
- Reduce (*Treat*) risks. This is achieved by undertaking surveys, redesign, use of other materials, use of different methods or by changes in the procurement plan. The approach is less drastic and more common than elimination. Reduction is accomplished by actions undertaken by members of the project delivery team. For example, the design team can change the design to eliminate that bit of the design which causes the risk. Such change might result in increased based cost. The team can undertake surveys to provide better information thus removing uncertainty and thereby reducing the risk. Surveys will cost money to conduct. The team members can review different materials or equipment to achieve the required outcome. By adopting a more proven method or solution (which by definition is less innovative) the risk is likely to be reduced. They could propose different ways of working towards the project objectives and reduce risk in that way. They could package the work in different ways to reduce the interfaces between the trade contractors thereby reducing the risk. The effect of reducing or treating risks is generally to increase the base cost and reduce the risk allowance because the team is paying to undertake the risk reduction process.
- Insure (*Transfer*) risks. Some risks can be insured or transferred to other parties. Insurance is a straightforward transaction where, by the party insured paying a premium, should the risk occur the insurance company will recompense the damage caused. Another commonly used way of transfer is to allocate risk to the contractor through the contract. Allocating the risk in this manner does not actually reduce the risk. It normally results in the client paying a premium to transfer risk to the contracted party. Adopting this strategy results in the client paying a premium to reduce his or her risk. There is a common misconception that transferring a risk to another party in the project delivery supply chain eliminates the risk. This is, unfortunately, not the case. The contractor or other party to whom the risk is transferred must still manage the risk. The risk is still present. Should

they fail to eliminate, reduce, insure or contain it, they must pay the consequences. If this is not done properly, the impact of the risk can still ultimately revert to the client.

- Contain (*Tolerate*) the risk within the unallocated contingency. This is the strategy to adopt for all minor risks. It involves no active management. The team judge that, should the risk occur, the cost is affordable within the risk allowance or contingency fund allowed for the project. Because these risks are less severe the competent management team will feel comfortable about addressing them only if they occur. Only then will they have an impact. However, if they occur, the full impact of the risk must be borne by the party to whom that category of risk is allocated through the contract.

Allocating management actions

Which strategy to adopt for the risks that have been identified will depend upon their severity (the product of likelihood and impact) and the team's 'risk appetite'. For each risk it is necessary to consider who is accountable should that risk occur. This person is normally called 'the account owner', and will be a senior manager or board member. The team must also decide who can best manage the risk, either on his own or in collaboration with others. This person is normally called 'the action owner'. Next the team needs to consider what the action owner can do to implement one of the strategies outlined above. This will be 'the management action'. Finally the team needs to decide by when the action should be completed and when it should be reviewed.

In defining the action that the action owner should take it is necessary to keep things in proportion: to assess the resources needed to undertake the action and compare these with the impact should the risk occur. There is little point in expending more resource to manage a risk than the cost of its impact should it occur.

If the project team is to maximise value and control risk, it is necessary to have effective value and risk management implementation plans in place. Essentially these should show individual responsibilities, how they communicate with one another, how proposals to improve value or reduce risk will be implemented, by whom and within what timescales. The plans will also set out the timetable for regular reviews and formal studies.

Value and risk management studies should be planned from the outset of any project. They should not be regarded as sticking plaster remedies to dig the team out of a hole when things have gone wrong. Sadly, this is often the way they are used, with the result that so-called value management exercises become little more than emergency cost cutting and risk studies are reduced to fire-fighting plans. The earlier that risk and value management are started, the more benefits will accrue. Figure 6.1 identifies the potential to add value at the start of a project and the increased risk of making a change during construction.

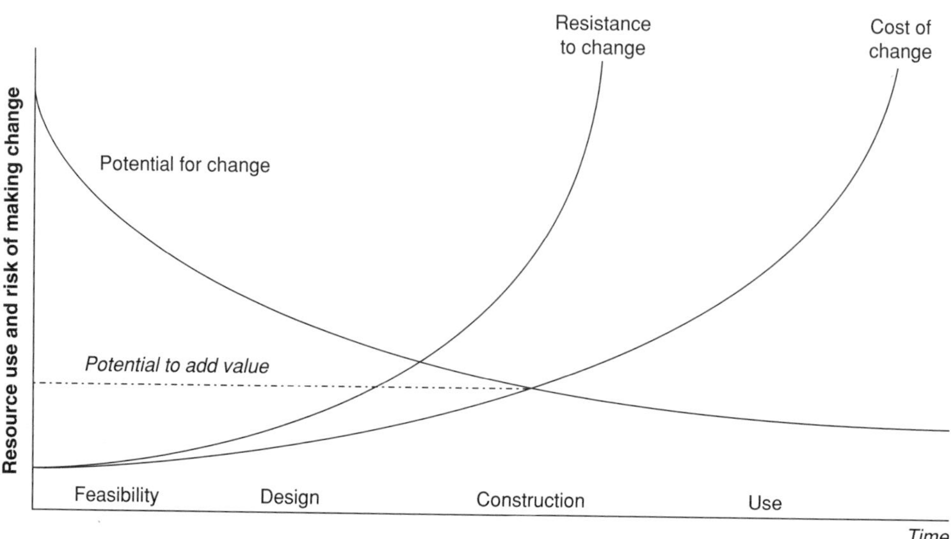

Figure 6.1 Opportunities reduce with time.

Value and risk are complementary

Both risk and value management are needed to maximise the chances of project success. The reason for this lies in the different but complementary objectives of each discipline outlined above. Value is maximised using value management. Uncertainty, and consequent value destruction, is minimised using risk management.

Similarities in the processes

Whilst the processes may differ in detail, they have the following similarities:

- the preparation stage, to understand the project and the issues relating to it
- the requirement for consultation with and involvement of the main stakeholders
- the use of facilitated workshops involving a balance of stakeholders, disciplines and characters
- the development of value and risk profiles by which progress towards improving value and reducing risk may be assessed
- the development of proposals to improve the project, and management actions to implement them
- the need for an explicit implementation plan
- the written record, or report on the outcome, providing a clear audit trail
- the need for regular reviews to monitor implementation and report progress.

The integrated process

Combining the two processes within a single programme of studies is therefore logical and practical. Essentially the integrated process comprises a number of formal studies that coincide with key milestones, or decision gateways, throughout the life of the project.

Between the formal studies, the progress of implementation and management actions should be reviewed by a responsible person on a regular basis and reported in the regular project reports. These reviews and progress reports are likely to be conducted and reported separately. This is because different people within the project team may be responsible for conducting them.

This process is illustrated in Figure 6.2.

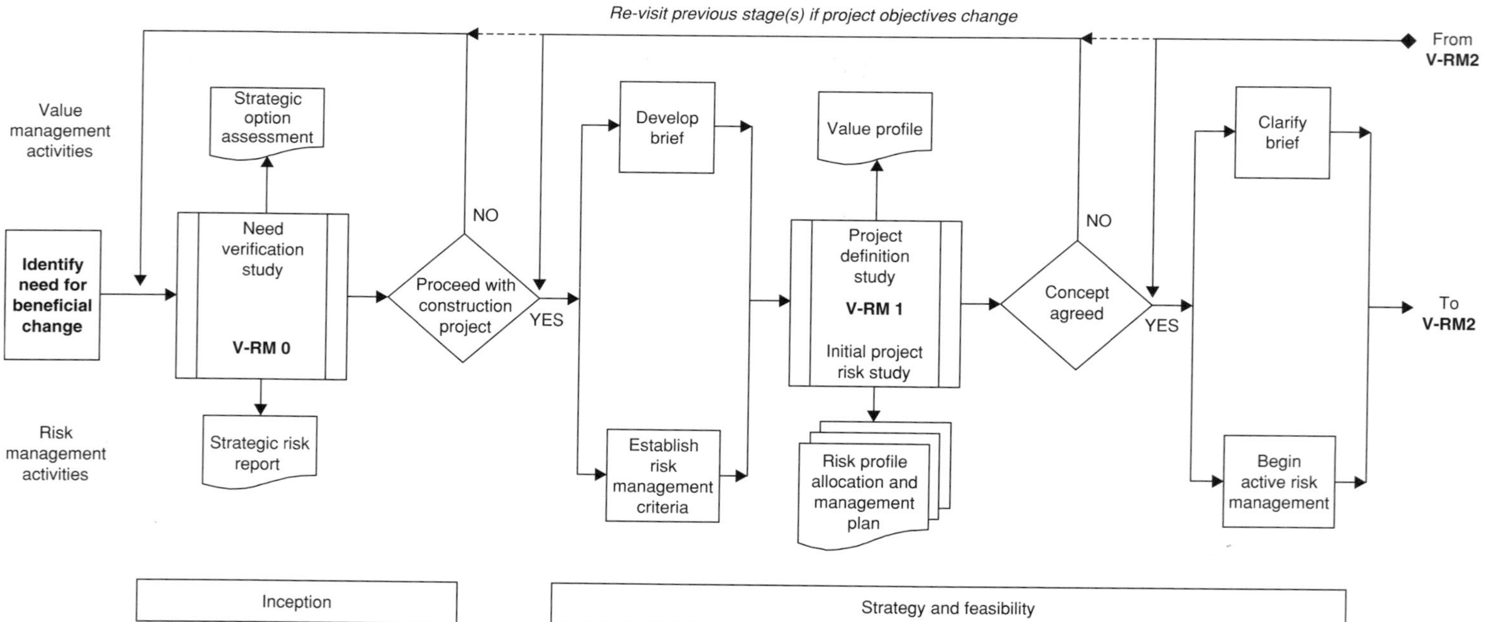

Figure 6.2 The integrated process of risk and value management.

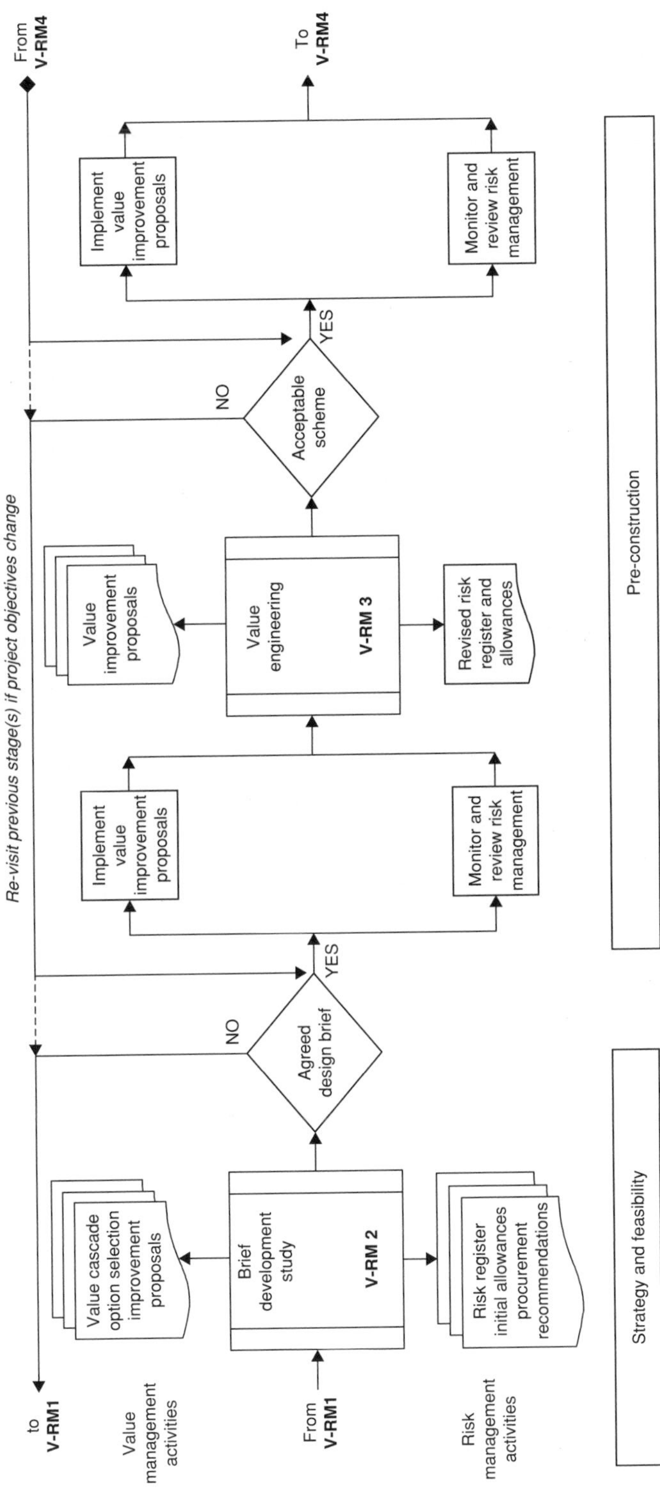

Figure 6.2 (*Cont.*)

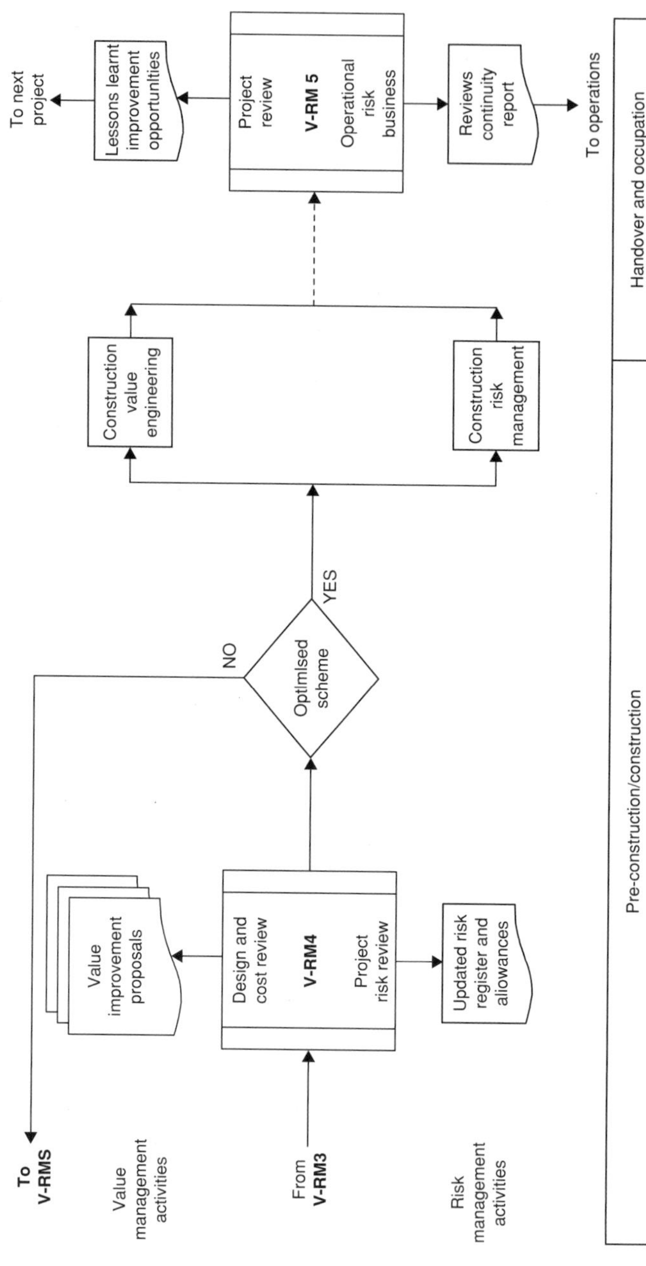

Figure 6.2 *(Cont.)*

Chapter 7

Fixed Price and Cost Reimbursement

Fixed price and cost reimbursement are terms used for convenience to define the two methods of making payment. However, their use is misleading because, although they indicate the principal methods available for paying for work done, very rarely is any contract discharged entirely by one or other method. Usually a combination

... considering the many combinations ...

of both methods is employed and the contract is named according to that which predominates.

The principles, however, remain valid. Fixed price items may be defined as items paid for on the basis of a predetermined estimate of the cost of the work, including an allowance for the risk involved and the market situation in relation to the contractor's workload. The estimated price is paid by the client, irrespective of the cost incurred by the builder.

Cost reimbursement items may be defined as items paid for on the basis of the actual cost of the work.

The JCT Prime Cost Building Contract (2005) for building work offers a detailed definition of actual cost to the builder under the heading 'Definition of Prime Cost' in the second schedule.

Fixed price

It will be evident, from the definition of a fixed price item, that fixed price can apply to a unit rate, a section of work or equally to a complete contract. Similarly, it must be appreciated that a contract may consist of a multiplicity of unit rates, a series of elemental or trade sections or a single lump sum. Understanding this should dispose of the common misconception that a fixed price contract is a lump sum contract.

A fixed price contract need not have a finite sum attached to it at the beginning of the contract. A schedule of fixed rates (without quantities) is a fixed price contract, because the basis of payment has been predetermined – the price being fixed, only the quantity of work is unknown and this is ascertained by measurement as the work is done. This is still true even if the rate varies with the quantity done or used.

Cost reimbursement

In this system payment is not based on a predetermined contractual estimate of cost. The contractor is paid whatever the work actually costs him within the limits of the contractual arrangement, which will lay down strict rules or formulae for the ascertainment or calculation of that cost.

The essential difference between fixed price and cost reimbursement is that in the case of the fixed price contract the contractors undertake to do the work at a price they have estimated in advance. If they are incorrect in their estimation, then they take the risk for being wrong. In a cost reimbursement contract, however, the employer pays the actual cost to the contractor, often described as an 'open-book' approach. If that cost is higher or lower than any estimates which may have been given for the project before it was started then the employer automatically pays for the extra or gains from the savings.

Application to contract elements

Having established the principles, it is now possible to consider, in practical terms, the differing elements that make up a contract and how the fixed price or cost reimbursement principles apply in each case.

Unless the contract is for a single lump sum figure to supply a specified building, a 'fixed price contract' is usually made up of several elements, as follows:

- preliminaries – often a series of individual sums to represent parts of the project, such as site management costs, plant, etc.
- unit rates – work fixed in place
- PC sums
- provisional sums – work anticipated to be required but not yet designed, including contingencies
- provisional sums – for work executed by a local authority or by a statutory undertaker executing work solely in pursuance of its statutory obligations
- profit – sometimes included in preliminaries and unit rates but sometimes separated as a sum or percentage in the summary.

At once it will be seen that the first two items are of a fixed price nature while the third embodies the cost reimbursement principle. Provisional sums are a way of expressing that part of the employer's budget, which is not fully defined, in a contract and can eventually be resolved either way, and sums for work by local authorities and statutory undertakers are frequently on a cost reimbursement basis. Profit, however included, is always on a fixed price basis.

Even in a cost reimbursement or 'cost plus' contract the allowance included for overheads and profit is committed in advance as a lump sum, or percentage of the cost spent on the labour and materials, regardless of whether the actual cost of the overheads and management is higher or lower than this. There are therefore cost reimbursement items for labour, materials and plant, but a fixed price item for management. Where a fixed lump sum fee is given for management, an estimate of cost will be made in advance, perhaps as a percentage, as a basis for calculating this fee.

In practice, as explained above, so-called reimbursement contracts frequently contain elements of fixed price and many fixed price contracts contain elements of cost reimbursement.

Fluctuations

An element of cost reimbursement that may come into otherwise fixed price contracts concerns the payment for fluctuations in wages and materials. A distinction is frequently made here by calling a contract where fluctuations are not paid a 'firm price contract'. When fluctuation clauses are included in the contract and an adjustment is made for the increase or decrease in the basic cost of labour and materials, that part of the contract is dealt with on a cost reimbursement basis where the payment is based on the market price fluctuation. However, when the payment for increases or decreases in labour and materials is based on a formula related to a national index, contractors will be reimbursed at a level which may or may not reflect the price fluctuations which they encounter.

Target cost contracts

Having defined fixed price and cost reimbursement and indicated that they are the principal methods of payment, that is not the end of the matter. It is possible to combine them, not only by having part of the work paid for on one basis and part on the other, but by applying both principles to the same work. In fixed price contracts, contractors take the risk of their estimates being wrong, while the reverse is the case with the cost reimbursement contract where the employer is at risk in relation to the final cost. If, however, it is desired that the risk should be split between the two parties, then payment can be made in such a way as to allow this. A target cost contract is such a method and the system is explained in detail in Chapter 10.

Use

Just as, in discussing competitive and negotiated contracts, it was suggested that negotiation should be used only where it can be shown to be more advantageous to the employer than competition, so a similar principle applies here: fixed price will, in the majority of circumstances, benefit the employer.

The appropriate allocation of risk between the employer and the contractor will determine the apportionment of fixed price and cost reimbursement in a contract. In assessing the situation there are two main considerations, namely the circumstances of the employer and of the contractor.

The employer's position

The following criteria should be considered:

- the employer's financial position
- the employer's time requirement for building work
- the employer's corporate restrictions
- the building market.

If employers have a defined budget and adequate time for their professional advisers to obtain fixed price quotations for the work, they should pursue that route in order to minimise their risk. If an employer has only £100,000 to spend, the advisers should limit the risk, so that this figure is not exceeded. However, if time is more important than money, the employer's position may be best protected by taking more financial risk and moving towards a cost reimbursement type of contract. It is important that the employer is advised on the correct balance between the two methods of payment. Any attempt to reduce the financial risk without providing adequate information to the contractor will undoubtedly backfire, with the employer having to pay more as the contractor will price for all eventualities to cover potential risk which may not eventually arise.

Employers may have a predetermined method of procurement for building work which will override their best interests. For example a local authority may require fixed price tenders to be obtained from an accountability point of view, yet this method may preclude the authority from having the work carried out during a particular financial year and so may lose the money allocated for the scheme.

When work in the building industry is scarce contractors will inevitably be prepared to accept greater financial risk in order to obtain work, and the employer's professional consultants should advise on that basis.

The contractor's position

The following criteria should be considered:

- the contractor's financial position
- the contractor's financially acceptable risk level
- the extent to which the work is defined
- the building market.

As with the employer who has limited financial resources, contractors in a similar situation cannot afford to take financial risks. Therefore, when asked to take such a risk, they will add a sufficient amount to their fixed price to cover the risk. In certain situations they may not be prepared to take the risk at all and will refuse to tender for the work. On the other hand, contractors with substantial financial resources are in a position to take on more financial risks and submit fixed price tenders without prejudicing their immediate financial stability. Obviously, they must be satisfied that the risks they are prepared to take in pricing will even themselves out over the whole range of work that they undertake.

The financial risk level that the contractor is prepared to take is a commercial decision. Two contractors with similar financial resources may have a different tolerance level for risk. For example, one may be prepared to estimate for material and labour cost increases, whereas the other will require to be reimbursed for any such increases. This action may preclude the latter from being invited to tender.

Where it is difficult, or impossible, to define the work clearly, as in some alteration and refurbishment schemes, it can be beneficial for at least some of the work to be let on a cost reimbursement basis. Attempting to estimate a fixed price for work that is not properly defined will result in the items being heavily priced to cover the risk. In addition, an undefined global item could be contractually unenforceable if it proves to be priced inadequately, and conversely the employer will find it difficult to reclaim any money if the work involved is less than the undefined item allows for.

Where new building techniques are being used, it may be in the interest of both parties to have the work carried out on a cost reimbursement basis, as the true risk may not be definable. Attempting to allocate a fixed price may, in these circumstances, result in too low or too high a price – with disastrous effects all round.

Market conditions can affect a contractor's attitude to tendering. When the building market is depressed, the contractor may take the view that he is prepared to take greater risks and tender on fixed prices. In a more buoyant market, he would be more likely to look for a cost reimbursement basis.

Programme

It will be appreciated, from what has already been stated, that cost reimbursement contracts are most commonly used where completion in the shortest time is essential, or where it is impractical to define the work fully before it is executed.

Figure 7.1 shows a diagrammatic representation of the different sequence of events between fixed price and cost reimbursement contracts. The vertical divisions of the matrix are not intended to indicate any particular time interval – they are provided purely to assist visual comparison of the alternatives. Neither is the length of the activity bars necessarily significant, except in allowing a comparison to be drawn between the two systems. In practice, the duration of any activity must be agreed between the participants, so as to achieve the employer's expectations.

However, Figure 7.1 indicates that a considerable time saving can be achieved using a cost reimbursement contract – not only in achieving completion of the work, but also in completing the final account. Since the cost verification process has to be commenced from the very beginning of the contract, if progress payments and cost reporting are going to be realistic, the final account should almost be complete when payment is calculated for the last period of work.

Summary

When there is sufficient time it is beneficial to carry out investigative work and produce as much detailed design and specification work as possible. This work will

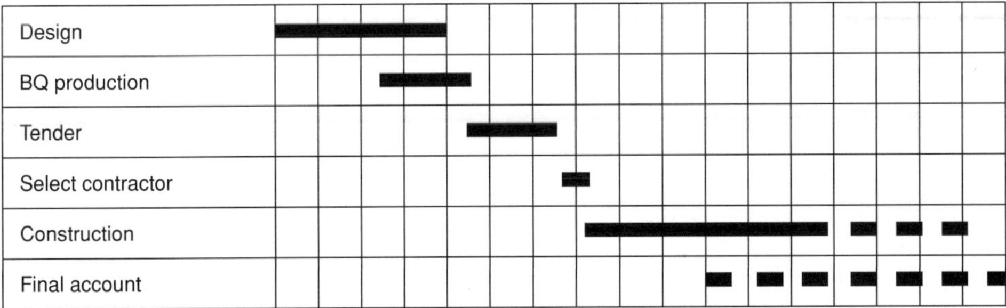

Cost reimbursement (negotiated tender)

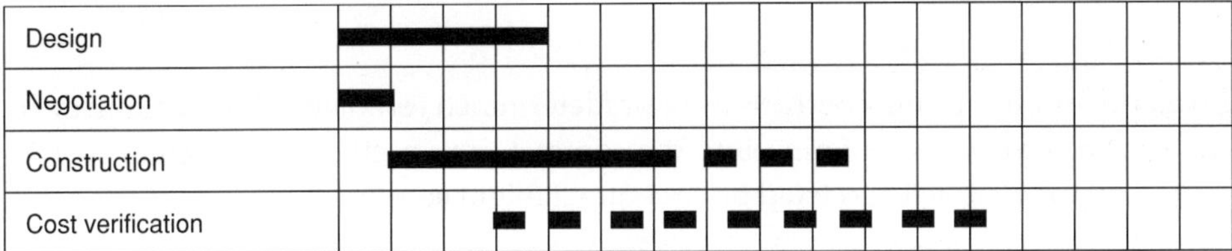

Figure 7.1 Sequence of events – fixed price and cost reimbursement.

reduce the risk element in the contract and influence the type of contract eventually used. On all contracts, payment is made either by fixed price or cost reimbursement: FP + CR = 100%. Almost all building contracts are a combination of both.

With the increase in use of cost reimbursement and management type contracts, often let on an 'open-book' basis, it is clear that employers are prepared to take more financial risk in pursuit of shorter procurement times and lower prices. Partnering arrangements are often set up on an 'open-book' basis – with a view to all parties achieving best value and sharing the financial risks and benefits more equally than is the case with traditional fixed price contracts.

Chapter 8
Fixed Price Contracts

Fixed price or lump sum contracts are a predetermined estimate of the work made by the contractor prior to entering into the contract, irrespective of the actual cost incurred by the contractor. A fixed price contract might comprise:

- a single lump sum
- a series of element or trade totals
- a multiplicity of unit rates.

Historically, fixed price contracts have accounted for the majority of building contracts and are still considered as being the norm, by number of contracts if not by value. Any departure from a fixed price contract, in favour of cost reimbursement or target cost contracts, is only recommended when the circumstances specifically require a different approach in order that the best value for money can be obtained from the employer's perspective.

The majority of the JCT forms of contract are essentially fixed price contracts, albeit that they usually contain optional clauses to allow an element of reimbursement of fluctuation costs, should fluctuation be stated in the Contract Particulars as applying. Optional fluctuation clauses in JCT contracts pass the risk of inflation for labour, materials, tax, levies and fuels from the contractor to the employer, but other major areas of risk, such as management, supervision, resource availability and productivity, remain at risk to the contractor.

JCT fixed price contracts

The main JCT fixed price contracts together with the other items necessary to complete the contract documentation are set out in Table 8.1.

Table 8.1 JCT Forms of Contract.

Form of contract		**Other documentation**
(1) Standard Building Contract		
(a) with quantities		drawings bills of quantities (incorporating the specification)
(b) with approximate quantities		drawings bills of approximate quantities (incorporating the specification)
(c) without quantities		drawings specification
	optional	work schedules contract sum analysis/schedule of rates
with contractor's designed portion (CDP) (for optional use with quantities, approximate quantities and without quantities)	*additionally*	employer's requirements contractor's proposals CDP analysis
(2) Design and Build Contract		employer's requirements contractor's proposals contract sum analysis
(3) Major Project Construction Contract		the requirements the proposals
(4) Intermediate Building Contract		
(a) with quantities		drawings bills of quantities
(b) with approximate quantities		drawings with or without specification work schedules (priced)
	or	specification (priced) with or without work schedules
	or	work schedules with contract sum analysis or schedule of rates
(5) Intermediate Building Contract With Contractor's Design		As the Intermediate Building Contract
	additionally	employer's requirements contractor's proposals CDP analysis
(6) Minor Works Building Contract		drawings specification work schedules schedule of rates (priced)
	or	drawings specification (priced) work schedules
	or	drawings specification work schedules (priced)
(7) Minor Works Building Contract With Contractor's Design		As the Minor Works Building Contract
	additionally	employer's requirements

The Standard Building Contract

The SBC in its various versions is the JCT's most comprehensive statement on contract conditions. The forms contain detailed conditions regulating the rights and obligations of the parties, the powers and duties of the architect contract administrator and the quantity surveyor, and procedures appropriate to the variety of situations to be met on most projects of any size and complexity – even to the extent of covering the outbreak of hostilities, war damage and terrorism.

The With Quantities, Approximate Quantities and Without Quantities versions all contain similar details and differ only in respect of those conditions referring to the contract bills, remeasurement (in the case of approximate quantities) and schedule of rates (where quantities are not measured).

The SBC form, derived from a long list of predecessors, has built up a considerable body of case law which helps clarify the manner in which the documentation is to be interpreted.

The appropriate version is selected and is dependent on the amount of information available at tender stage. It should be noted that with Standard Method of Measurement of Building Works 7th Edition (published by the RICS and the Construction Confederation) (SMM7) and Co-ordinated Project Information (CPI) the choice of form used is more directly related to the information available. Bills of quantities cannot, at least in theory, be produced without the appropriate drawings and specification information being available.

Recognising the concept of contractor's design, the SBC provides for a contractor's designed portion (CDP), where the majority of the design is produced by consultants and part is undertaken by the contractor. If a CDP is used, the basic documentation of the SBC is reinforced by the design obligations contained in the Design and Build (DB) contract, namely the Employer's Requirements referring to that portion of the works to be designed by the contractor, the Contractor' Proposals and the Contract Sum Analysis.

A comprehensive set of tender and contract documentation is available for subcontractors. This is fully co-ordinated with the main form of contract which ensures that the relevant conditions are fully compatible.

Provision is also made for reimbursing fluctuations in the prices of labour and materials, either by traditional means or by using the formula method when quantities are provided.

Design and Build Contract

In addition to the traditional fixed price contracts, there is another form worthy of note, the DB Contract. This form is different from traditional forms since its concept is based on a different premise, namely design provided by the contractor rather than that ordinarily provided by the employer's consultants. It is one of only two forms produced by the JCT which provides for all of the design to be undertaken by the contractor. The concept of the contractor providing the design requires alternative contract documents to those used in traditional procurement. These alternative documents comprise the Employer's Requirements, the Contractor's Proposals and a Contract Sum Analysis (see Chapter 12).

Major Project Construction Contract

The Major Project Construction Contract (MP) is one of the newer JCT standard forms and was originally drafted under the guise of the Major Project Form in 2003. Guidance to the MP advises that this form is for employers who have in-house contractual procedures and regularly undertake projects. The contractual provisions are therefore somewhat shorter than those of the SBC. The form is similar to that of the DB Contract and utilises requirements which are, in principle, the same as Employer's Requirements in the DB contract. However, the form does not preclude the use of traditional procurement. The requirements may simply comprise drawings and a bill of quantities but this is unlikely as those who are envisaged as using the contract are also likely to seek to transfer the design risk to the contractor.

The MP is sensible in its approach to the responsibility of sub-contractors. If a sub-contractor of the employer's choice is required to undertake part of the project, the contractor assumes full responsibility for the sub-contractor without systematically jumping through the hoops required under the naming provisions of the Intermediate Building Contract (IC) or the nomination procedure formerly practised under the SBC's predecessor, the JCT 98. In allowing the employer to name a sub-contractor and ensure that the contractor is responsible for the sub-contractor, the requirements simply need state so.

As with the employer requiring the contractor to be responsible for sub-contractors, the form also envisages that the employer's consultants, who were appointed to draft the requirements, might be novated to the contractor. However, the form does not set out the terms on which novation must take place, but merely provides the machinery for it. Therefore, it is prudent to select a standard form and state this as part of the requirements so that the contractor is aware of the terms and has the opportunity to price the risk before the contract is entered into.

The MP has a novel but, nevertheless, practical approach to completing a project for the mutual benefit of both parties. The form provides for:

- Accelerating the project to practical completion before the completion date.
- Bonus payments should practical completion occur before the completion date.
- Cost savings and value improvements for which a benefit may arise in the form of:
  - — a reduction in the cost of the project;
  - — a reduction in the life cycle costs associated with the project; and
  - — any other financial saving to the employer.
- Novation of the design team.

If there is a disadvantage in using this form over the SBC or DB it is that, unless the employer is experienced in procuring projects and can provide much of the input as to the requirements, the additional cost of professional fees ordinarily incurred is likely to be above that of other JCT forms. In contrast to the SBC and DB forms, the MP is more facilitative in its approach than comprehensive and is for use by employers with experience and precedent to draw upon.

Intermediate Building Contract

The Intermediate Building Contract (IC) is produced in two formats. The first format is based on the premise that all of the design is completed by the employer's design team, and the second envisages that the contractor is required to complete part of the design. Essentially the main difference between the two documents is that the Intermediate Building Contract With Contractor's Design (IC/D) allows for contractor's design by containing a CDP in much the same way as this is an option with the SBC in its three versions.

To give a degree of flexibility the form contains a number of alternative clauses in the recitals and contract particulars which, depending upon the particular circumstances of the project, are selected (deleting those not required) to produce a with quantities or without quantities format. The form contains conditions that are shorter and somewhat less detailed than those of the SBC variants but more detailed than those of the Minor Works (MW) Contract. The IC Contract Guide suggests that the form is primarily intended for use where proposed works are:

- of simple content involving the normally recognised basic trades and skills of the industry
- without any building service installations of a complex nature, or other complex specialist work
- designed by or on behalf of the employer (although IC/D caters for an element of contractor design)
- defined adequately in terms of quality and quantity in the bills of quantities, specifications or work schedules.

The provisions for the reimbursement of fluctuations in the prices of labour and materials are similar to the SBC, albeit shorter, allowing for the traditional or formula methods as preferred (the formula method only being available when quantities are provided).

The IC provides employers with the opportunity systematically to select individual sub-contractors to undertake the works. These are identified within the contract as named sub-contractors. The advantage of naming sub-contractors is that they can be selected in advance of the main contractor, should specialist work be required; once a sub-contract is entered into, the named sub-contractor effectively becomes a domestic sub-contractor and the main contractor assumes total responsibility for the work (subject to various special conditions if the sub-contract has to be determined). Special tender and sub-contract forms are provided to administer this process to ensure compatibility with the main contract. It is important that this procedure is carried out in advance of the signing of the main contract as any sub-contract must be executed within 21 days of executing the main contract.

In this form there is authority for a clerk of works to give a 'direction' to the contractor but there is no provision for excluding people from the site. The only duty of the clerk of works is to act solely as inspector on behalf of the employer, under the directions of the architect/contract administrator.

Minor Works Building Contract

The Minor Works Building Contract (MW) is the simplest JCT form available and, like the IC, is available in two formats; one in which all design work is completed by consultants on behalf of the employer and the other in which the contractor undertakes to complete part of the design (MWD). The form is drafted for use with drawings and/or specifications and/or a work schedule, but no provision is made for the use of bills of quantities. MW guidance suggests that it is appropriate to use the form where:

- the work involved is simple in character;
- the work is designed on behalf of the employer (except the contractor's designed portion in MWD); and
- a contract administrator is to administer the conditions.

Although minor works might not be considered to extend over a long period of time, provision is made for fluctuations in the prices of labour and materials due to contribution, levy and tax changes.

Specialist sub-contractors can be incorporated by naming them in the tender documents or in subsequent instructions but there is no provision in the form for any special treatment or control of such specialists, nor are there any standard forms of sub-contract available for the use in that situation. The contractor does, however, remain wholly responsible for any works undertaken by a sub-contractor.

With regard to the provisions of other JCT forms, there are a number of significant omissions. These comprise:

- any provisions for insuring the employer's liability for damage to property other than the works, or for all-risks cover
- any provision for the ownership of materials on site to pass to the employer on payment
- any provision for opening-up and testing.

In respect of the first point above, there are provisions in the contract for the contractor to indemnify the employer against any expense, liability, loss, claim or proceedings for damage to any property. The ability to comply with this obligation, however, is dependent upon the solvency of the contractor.

As with the IC, or any other contract for that matter, any limitations could be overcome by drafting special instructions, but again there seems little point in doing so when there are more comprehensive forms available covering the omissions in this very simple form.

Advantages and disadvantages of fixed price contracts

The following points may be found useful in deciding to recommend to a client the use of a fixed price or cost reimbursement contract, and which of the various JCT forms of contract to use.

Advantages

- Depending on which elements of the fluctuations clauses are deleted the employer avoids the risk of variations in cost due to management deficiencies, shortage of resources, reduced productivity and, to some extent, inflation.
- Unless approximate quantities are provided the contract sum will define the employer's financial commitment, subject to contingency and any other provisional allowances and, of course, to any variations in the brief.
- There is an in-built incentive for the contractor to manage the work effectively and to complete the work as quickly as possible so as to maximise his profit.

Disadvantages

- The premium paid to the contractor for taking the risk is paid irrespective of whether or not the risk materialises.
- Time and production cost savings and improved productivity deriving from the learning curve are lost to the client; only the contractor benefits. This may be particularly significant on very repetitive jobs.

The choice between the various fixed price contracts available will depend on the other documentation available and the scope and complexity of the works.

Chapter 9
Cost Reimbursement Contracts

In Chapter 7, cost reimbursement was defined in terms of payment on the basis of the actual cost incurred by the contractor, irrespective of any estimate which may have been calculated for budgeting purposes or submitted as part of a tender. A cost reimbursement contract is therefore one in which the contractor is reimbursed actual cost plus a fee to cover overheads and profit.

By their very nature cost reimbursement contracts cannot have a finite sum either at contract stage or when a contract is signed. The contract sum will only be ascertained at completion when the final account is settled.

Since the contractor is reimbursed the actual cost, it will be seen that, with this type of contract, the employer carries all the risk – inflation, management efficiency, effective supervision, resource availability and productivity. The contractor does take on a contractual obligation to carry out the work as economically as possible, having regard to the nature of the works, the prices of materials and goods and the rates of wages current at the time that the work is carried out, together with other relevant circumstances. It is nevertheless important that employers and their consultants have confidence in the competence of the contractor and that some way of controlling the contractor's method of operation is included in the contract documentation.

The fee

The fee paid to the contractor can be a percentage figure (applied to the prime cost) or a fixed fee (a lump sum figure). A fixed fee has the advantage of providing an incentive for contractors to work efficiently (to maximise their return). A percentage fee lacks this incentive and is sometimes considered to be an open cheque for contractors – the more they spend, the higher their fee – and is very much akin to the concept of dayworks. Although this may be considered an extreme view, it does serve to highlight the danger of not providing an incentive for contractors to work efficiently. However, if the scope of the work is not sufficiently defined, a fixed fee may not be acceptable to any contractor and a percentage may have to be agreed.

A refinement of the fixed fee method is to incorporate a provision for the fee to be varied as the final estimate of prime cost varies in relation to the original estimate of prime cost. This helps in dealing with variations.

As a further incentive to efficiency, it is possible, with either a percentage or fixed fee basis, to incorporate a provision that the fee is only adjusted if the final estimated prime cost varies from the original estimated prime cost by more than 10%, or some other agreed percentage. This adjustment could apply to both upward and downward movement.

The Prime Cost building contract

The Prime Cost building contract is one of only two JCT cost reimbursement contracts. The other is the Measured Term contract and is discussed in Chapter 13. The fact that there are only two JCT forms possibly indicates that only a minority of contracts are let on this basis. It also adds weight to the view, expressed in Chapter 8, that fixed price contracts may be considered to be the norm and that cost reimbursement contracts should only be used on those projects where specific conditions render them more suitable, e.g. emergency repairs, investigatory works and such other cases where it is not possible to identify in advance the scope of the work required to be done.

Nevertheless the advent of partnering arrangements has led to an increase in the use of this form of contract. If it is properly administered by both the contractor and the architect/contract administrator, the employer may benefit from not paying for risk items that do not materialise.

Characteristics of the form

The main feature distinguishing this form from that of fixed price contracts is the agreement for payment to the contractor. Payment is based on the prime cost of the work, as defined in the documentation, and a fixed or percentage fee.

The form contains detailed conditions regulating the rights and obligations of the parties, the powers and duties of the architect/contract administrator and the quantity surveyor, and procedures appropriate for the administration of the work.

By its very essence a cost reimbursement contract does not make any provision for reimbursement fluctuations, these are dealt with automatically as invoices and time sheets are priced at rates current at the time the work is carried out.

Advantages and disadvantages

Since fixed price contracts are considered suitable for most circumstances, there should be cogent reasons for recommending to a client the use of a cost reimbursement contract. Consideration should be given to the advantages and disadvantages of such a contract when evaluating whether or not the use would be to a client's advantage in any particular situation.

Advantages

- Work can be commenced on site immediately provided that sufficient specification information is available.
- It is often the only way of carrying out investigation work.
- It is ideal for coping with emergencies such as making buildings safe after fire damage.
- Production cost savings and improved productivity can derive from the learning curve resulting from the repetitive nature of jobs, to the benefit of the client.
- The client pays no premium to cover risk – he pays only the actual cost of whatever is deemed to be necessary if and when those risks actually materialise.

Disadvantages

- The client's ultimate commitment is not known.
- There is little incentive to the contractor to employ his resources efficiently (but see section above on the fee).
- The client carries nearly all the risk on the contract.
- Checking the prime cost can be a very complicated and expensive operation.

Budget and cost control

The procedure for creating a budget and controlling the cost in the early feasibility and design stages is similar to that for a normal fixed price contract. An early budget estimate should be prepared and there is no reason why cost planning should not go ahead normally. Furthermore, cost checking during the evolution of the design can also be carried out as with other types of contract.

It is important that the documents showing the estimated cost of the project should be as detailed as possible so that the work is clearly defined, and that the basis on which the fee is calculated is established, whether it be a fixed fee or a percentage fee. It is also necessary that the estimated prime cost should be divided between the work to be carried out by the contractor's own labour and the work to be carried out by sub-contractors.

If it is decided that the fixed fee should be settled in competition and the contractor selected in that way, the detailed estimate must be sent to the tendering contractors as one of the tender documents.

Administering the contract

The procedures for operating fixed price contracts (the norm) are dealt with in some detail in Part IV, Contract Administration. This chapter considers only those procedures relating to cost reimbursement contracts which differ from the norm.

It is not normally advisable to change clauses in standard forms of contract, but the Prime Cost Building Contract does contain a good deal of procedural matter which it is perfectly reasonable to vary in order that the most efficient method of working can be established on a particular project. For example, if joinery is likely to be supplied by the contractor's own workshop, it may be desirable that the calculation of prime cost for that work should be on a different basis to the prime cost of work on site. It is also permissible to consider the incidence of small tools, perhaps even consumables, and to judge whether it is likely that fewer disputes will arise if some of these items are included in the fee rather than the prime cost.

There are other matters of procedure in which the quantity surveyor will be particularly interested. For instance, for cost control purposes the contractor may be required to send in time sheets and invoices at regular intervals for checking. Often a procedure for checking deliveries and delivery notes is necessary. It is desirable to ensure that the contract covers the procedures required.

'... *a procedure regarding checking deliveries* ...'

Procedure for keeping prime costs

The definition of prime cost is crucial as generally what is not specifically included in the definition is automatically deemed to be included in the fee. With small contractors, who might be unused to working by this method, it is wise to point this out in some detail before the contract is signed so that there is complete understanding on both sides as to what is supposed to be included in the fee. This is the best way to avoid niggling disputes,

The quantity surveyor should agree with the contractor a working method of keeping the prime cost. Contractors are sometimes inclined to take the line that, as they have been chosen to carry out work on this basis, they should have a free hand and their method of keeping the prime cost should be accepted. However, a proper system of checking the prime cost is essential and does not indicate mistrust of the contractor: it is merely a prudent method of doing business that is only right and proper on accountability grounds. Furthermore, the contractor's internal method of keeping the prime cost may not, and in fact probably will not, coincide exactly with the definition of prime cost in the contract. The quantity surveyor will, however, be well advised to go along as far as possible with the contractor's normal costing system as this is certainly the most likely to produce error-free results.

There seems to be no reason why the quantity surveyor should not receive a copy of the weekly wage-sheet, particularly where this is prepared for the contractor in any case. If some different method is used, this must be discussed with the contractor. Quantity surveyors should not object to using another method, provided they get the same information as the contractor with regards to wages paid.

Labour resources

It is perhaps worthwhile at an early stage, just after the contractor is selected, to agree who is going to be the working full-time foreman to be paid as part of the prime cost, and which people are in a supervisory capacity and therefore come under the management fee.

A further fundamental point arising with cost reimbursement contracts is that, as the clients are taking the risk, they should have some control over the way the work is carried out. For this reason it is desirable that the extent of normal overtime to be worked is agreed in advance and that the contractor should then seek approval before working any further overtime. Thought should be given to all such matters, according to the type of project, and decisions made as to what should be incorporated in the documents for tendering.

Materials

A system must be worked out for the acquisition of materials. Generally speaking, competitive quotations should be obtained but it must be recognised that there may be occasions when materials are wanted so quickly that this is not possible. In such a case the quantity surveyor has to take a reasonable view and provided the materials bought are not above market price, there is no reason why they should not be included and paid for. Even where competitive quotations can be obtained, the quality of service likely to be given to the contractor should also be considered. It may be that the lowest quotation is not the most advantageous. Such instances are, however, likely to be rare.

Some system must be worked out to ensure that materials invoiced match up with materials delivered. This can be done by means of delivery tickets, and frequently contractors work such a system on their own fixed price contracts. An overall check on any particular material can be made by an assessment from the estimated prime cost. In many cases a physical check on site can easily be carried out.

Plant

The conditions of contract should state how plant is to be charged. The two main plant hire schedules in current use are:

- The RICS Schedule of Basic Plant Charges
- The Federation of Civil Engineering Contractors Daywork Schedule.

Although both these schedules are intended for use in connection with dayworks under contract, they can, with suitable adjustment, be used for cost reimbursement contracts. Generally, however, it is assumed that if the plant is wanted for a long period on site, it may be cheaper to buy it outright. It may be desirable therefore that some limit upon the period of hire of plant should be laid down in the contract. Sometimes this is expressed in terms of not paying more in hire than, say, 80% of the capital cost. At end of the contract, purchased plant can be sold or retained by the contractor, with an appropriate credit being included in the prime cost calculation

Plant hire will invariably bring with it the question of consumable stores and discussion should also take place early on how these will be treated.

Credits

A system should be established for allowing credits for surplus materials and also credits for salvage, such things as old lead taken in from existing buildings. Similarly, material supplied by the client must be taken into account when agreeing the management fee. The sale of purchased plant has already been referred to above.

Sub-letting

On all cost reimbursement contracts, there will be a proportion of work which is sub-let on a measured basis. This may be to a sub-contractor, named or otherwise, or possibly to a labour only sub-contractor. Provided the arrangement is economical and competitive, there is no reason why it should not be adopted. However, if the proportion of work being sub-let rises higher than that estimated when the fee was originally fixed, there may be a case for a variation of the fee. This will have to be established before the contract is signed and, for this reason, it is desirable that any contractor's intentions regarding sub-letting on this basis should be established before they are finally selected.

Defective work

Most prime cost contracts have a clause indicating that the contractor should make good defects at his own expense. This can raise problems. If the defects are made good after practical completion there is no difficulty in separating the costs involved, but if they are made good during the progress of the work then some specific separation of the cost must be made. A kind of negative daywork charge is needed; this is very difficult to apply in practice.

Cost control

Valuations for cost reimbursement contracts will normally be done on the basis of the labour paid and the invoices submitted by the contractor each month. The retention may be a percentage of the value of the work done or a proportion, say 25% of the fixed fee. There is always a considerable interval between work being carried out and invoices being submitted, and therefore, for cost control purposes, a specific reconciliation between the actual prime cost and the estimated prime cost must be made.

To do this properly one has to go back to the detailed estimate of prime cost used for tendering and establish what proportion of that work is done. Against this one sets the actual prime cost, plus the labour and material used but not yet invoiced. The difference between the two will show the extent to which the project is either saving on (or exceeding) the estimated prime cost. From this reconciliation estimates of the final cost of the project can be made. Obviously, in taking the original estimate of prime cost, allowance has to be made for any variations. Although cost control with cost reimbursement contracts is more difficult than with fixed price contracts, it must be carried out and should be reasonably satisfactory.

Final account

The final account will, of course, be the total of the actual prime cost plus the management fee. It may be that during interim valuations various items are put into a suspense account pending settlement as to whether they are a proper charge against prime cost or not. If possible, items in the suspense account should be settled and either included or rejected as soon as possible. The procedure for the final certificate, maintenance and handing over of drawings etc. in relation to the building is similar to that of normal fixed price contracts.

Chapter 10
Target Cost Contracts

Chapter 7 examined the relationship between fixed price and cost reimbursement contracts and established the position of the target cost contract as a half-way house between them. Target cost contracts are a refinement of ordinary cost reimbursement contracts and introduce an element of incentive for the contractor to operate efficiently by transferring a proportion of the risk to the contractor. These are sometimes referred to as 'gain-share pain-share' contracts.

The essential feature of the target cost contract is that the difference between the actual cost and the estimated cost is split in some way between the contractor and the employer. The philosophy behind this is that if the contractor cannot complete the work for the estimated cost it is not right for the employer to pay the whole actual cost, as at least some of the increase may be due to the inefficiency of the contractor rather than to inaccurate estimating. On the other hand, if the contractor completes the work at a lower cost than that estimated, it may be assumed that some of the decrease is due to his own efficient management and, therefore, some of the gain should go to him. In this way the target cost contract gives the contractor a built-in incentive to manage the work as efficiently as possible.

It also, of course, gives him an incentive to increase the estimated price as much as possible in the first place. It is therefore essential that employers take steps to ensure that their interests are safeguarded by employing expert advice in evaluating the estimated price. This may be negotiated with the contractor before work is started or can be established by competition as part of the original tendering process.

Target cost contracts, however, should not be entered into lightly. They are expensive to manage, involving as they do both accurate measurement and careful costing on the employer's behalf. This is no doubt one reason why target cost contracts are somewhat rare but it is important to be aware of the availability of the system should the need arise.

In a target cost contract, payment is made partly on an estimated fixed price basis and partly on the actual prime cost. This principle can best be understood by looking at two simple examples, one showing a saving on the original target and the other showing extra:

	£
Target (i.e. estimate total prime cost including allowance for head office overheads and profit)	100,000
Actual prime cost plus overheads etc.	90,000
Therefore saving on target	10,000
If saving is split on a 50:50 basis, payment to contractor becomes	95,000

Example 10.1 Target cost contract with saving.

Target (as before)	100,000
Actual prime cost plus overheads etc.	110,000
Therefore deficit between actual cost and target	10,000
If deficit is split on a 50:50 basis, payment to contractor becomes	105,000

Example 10.2 Target cost contract with overspend.

There are various points to consider in studying these examples:

- The figures above have been deliberately made simple: in practice the target must be revised to account for all the variations. It will, therefore, be the revised target which is compared with the actual cost to establish the saving or extra.
- The split can be in any proportions previously agreed. Thus, if it is desired that the contractor should take more of the risk, the proportions will be agreed so that the employer pays a figure nearer to the revised target and further from the actual prime cost.
- Various methods can be used to achieve a more sophisticated calculation by splitting the work between that of contractors themselves and that of their own and named sub-contractors. In addition, their head office overheads and profit can be covered by either a fixed fee or a percentage fee. In a case of a fixed fee, it would only alter if the revised target varied from the original.
- The target can be obtained in various ways. The importance of reasonable accuracy has already been stressed. The best way, undoubtedly, is by means of a full and accurate description of the whole, properly priced out and agreed between the contractor's and employer's quantity surveyors. There may, however, be circumstances when a bill of approximate quantities will suffice. For specialist work a method of estimating in common use by the particular trade, such as a labour and material bill, may be appropriate.

In the simple examples given, prime cost is assumed to be prime cost to the employer in accordance with the detailed terms of the contract defining prime cost. For

'... a more sophisticated calculation ...'

Columns A and B represent typical overspend and underspend situations

	Target	Actual	
		A	B
	£	£	£
(1) Prime cost of labour charges, clearly defining rates, allowances, fares, etc.	20,000	22,000	18,000
(2) Prime cost of materials and consumable stores, making allowance for discounts	25,000	26,000	24,000
(3) Sub-contractors work	20,000	21,000	19,000
(4) Site management (fixed)	12,000	12,000	12,000
(5) Plant, sheds, transport, etc. (fixed)	9000	9000	9000
(6) Office overheads, supervision and insurance (fixed)	8000	8000	8000
(7) Profit or fee (fixed)	6,000	6000	6000
	100,000	104,000	96,000
Column A – reduction in profit in respect of items (1) and (2) 50% of £3000		1500	
Cost to employer		102,500	
Column B – additional profit in respect of items (1) and (2) 50% of £3000			1500
Cost to employer			97,500

Example 10.3 Target cost contract with cost saving and overspend.

practical purposes, however, it may well be appropriate to negotiate some of the items on a fixed price basis. There are many possible variations and Example 10.3 illustrates one possible solution where three elements of the work involving site management (4), plant (5) and overheads (6) are on a fixed price basis, while being incorporated in an overall target cost contract. Sub-contractors (3) are included on a full cost reimbursement basis and, for the sake of simplicity, the profit or fee (7) is taken as fixed. This leaves the labour (1) and materials (2) elements (amounting to 45% of the total) subject to the target cost calculation.

Guaranteed maximum price contracts

This is a type of target cost contract where the agreed guaranteed maximum price cannot be exceeded.

Any saving on the guaranteed maximum price would be shared between the contractor and the employer in a similar fashion to Column B in Example 10.3.

Guaranteed maximum price contracts are very difficult to set up and administer for two fundamental reasons. First, it is difficult to agree a realistic maximum price, as a contractor will inevitably try to set the figure too high, making it meaningless, and second, all variations will require the maximum price to be reassessed.

Competition

Very often the target will be negotiated between quantity surveyors on both sides. However, it is possible for the target to be the subject of a competition between contractors. This particularly applies in civil engineering works. In the latter case the bills of quantities will form the tendering document on which the target is based. Normally the contractor submitting the lowest target is selected.

From then on, the procedure is the same as with a negotiated target. The work is measured and valued on the basis of the bills, any variations being taken into account. The contractor is paid on the basis of the actual prime cost plus the relevant fee, and the difference between that figure and the measured account is shared between the contractor and the client on the lines indicated earlier.

Contract

There is no standard form of contract available for target cost work. However, the contract presents no difficulties. The JCT Prime Cost Building Contract can be used for all but the payments clauses. It is essential to cover a definition of what is allowable in the prime cost and this, together with the procedure for keeping the prime cost, is dealt with in Chapter 9.

Advantages and disadvantages

The advantages and disadvantages of target cost contracts are very much in line with those of cost reimbursement contracts identified in Chapter 9. The only difference, and it is an important one, is that target cost contracts have the further advantage of including an incentive encouraging the contractor to operate as efficiently as possible. There is also an advantage in the flexibility of the system. The risk can be apportioned in different degrees depending on the circumstances of each party and their ability to take that risk. It need never be a black and white solution; all shades of grey are attainable according to the merits of each case.

The incorporation of a guaranteed maximum price (GMP) in a target cost contract introduces additional factors for evaluation. The advantage to the employer of the price ceiling has to be set against the disadvantages of a higher fee – for it must be appreciated that a higher premium will be charged for the additional risk taken by the contractor.

Use

The likely uses of target cost contracts have already been reviewed in Chapter 7, when considering the general use of cost reimbursement contracts, but, as indicated above, they do have the additional advantage of providing an incentive for the contractor to improve efficiency.

Chapter 11
Management and Construction Management Contracts

Management and construction management contracting are forms of contractual arrangement whereby the contractor is paid a fee to manage the building of a project on behalf of the employer. They are therefore contracts to manage, procure and supervise, rather than contracts to build.

Under management and construction management agreements, the contractor becomes a member of the employer's team – effectively the construction consultant – and it is from this factor that one of the main advantages of the arrangements is derived – the contractor working in harmony with the remainder of the construction team, instead of adversarially, as perceived in most traditional arrangements.

While management and construction management contracts are very similar in many respects, there is one essential distinguishing characteristic which is fundamental to the understanding of the two systems. This involves the contractual arrangements between the parties. In management contracting the works contractors, or package contractors, are in contract with the management contractor; in construction management, they are in contract with the employer. A diagrammatic representation of this difference is shown in Figure 11.1. It will be noted that, apart from this contractual link, the lines of communication are identical. This difference will be referred to again later, but the two systems can be evaluated together. For convenience, the term contractor/manager is used to denote the management contractor or construction manager.

Payment and cost control

To understand the way a management or construction management contract works, it is necessary to show how payment is made.

The building work is split into sub-contracts, generally referred to as packages. These are normally let in competition in the usual way, although some may be negotiated or let on a cost reimbursement basis. All discounts usually revert to the client. It should be possible to arrange the various sub-contracts so that all building work is covered, but in practice a small site gang will sometimes be required to unload and

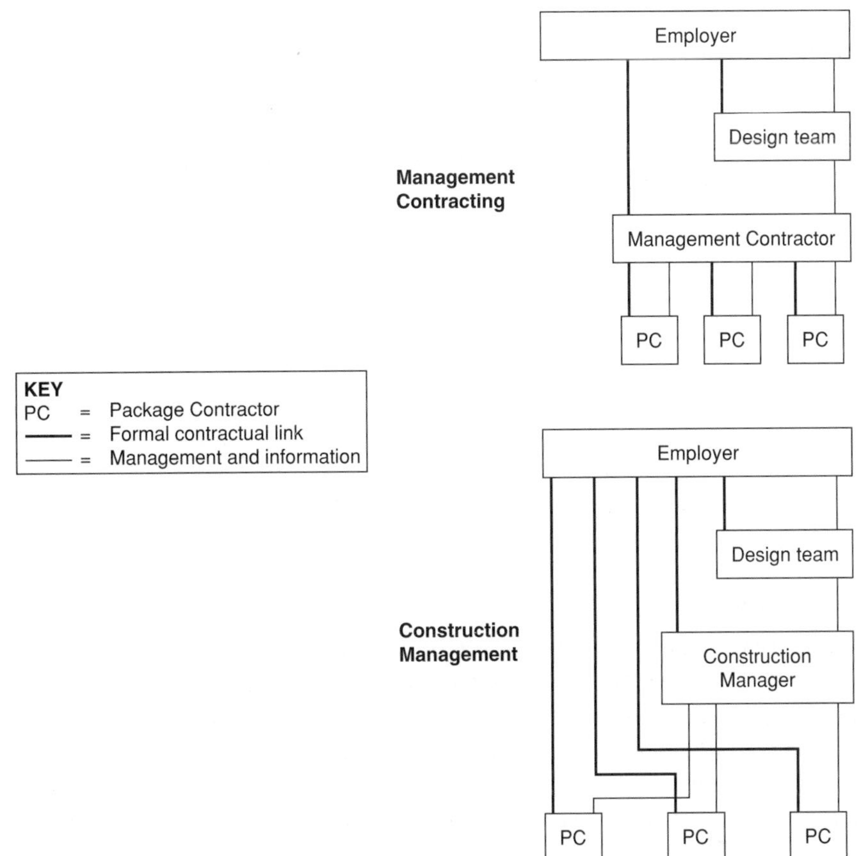

Figure 11.1 Differing contractual arrangements.

help with the movement of materials on site and general cleaning after trades. Apart from this it is unusual for management contractors to carry out any of the building work themselves, although there may be other companies within the management contractor's group wishing to tender.

The contractor/manager provides the site management team and such preliminary items as offices, canteen, hoardings, etc. which are usually reimbursed at cost. On some projects, however, the cost of these forms part of his tender and is a fixed lump sum. In either case, a schedule forming part of the contract clearly defines what is required.

In addition the contractor/manager will be paid a fee. This fee is in respect of overheads (i.e. head office charges) and profit, and is usually expressed as a percentage of the final prime cost of the works. It is therefore important to distinguish the site management team from the head office staff and facilities involved. Alternatively, the fee may be based on the cost plan and not be subject to change. Thus if the cost goes up, for no good reason, the effective percentage for overheads and profit on the actual cost decreases.

A cost plan will have been prepared prior to the appointment of the contractor/manager. As soon as the contractor/manager is appointed he must liaise with the quantity surveyor to prepare an estimate of the prime cost showing the estimated value of all the sub-contract packages, site management and other costs. This is then agreed by the client and architect as reflecting the required level of specification and becomes the estimate of prime cost for the project. Valuations and certificates for payment are prepared in the usual way. Cost reporting is carried out on a package by package basis against the estimated prime cost.

Some employers require the contractor/manager to be bound by a guaranteed maximum price (or GMP), this being a sum above which any expenditure will not be reimbursed. This, however, cannot be achieved until the design and specification of the building are fairly advanced and therefore, given the nature of these contracts, cannot usually be stated in the original contract documentation.

Selection and appointment of the contractor

Selection of the contractor/manager can either be on a negotiated basis or, as is more usually the case, in competition. Negotiation is easier than for a fixed price contract as the payment for most of the work relates to sub-contracts, which will be dealt with separately after selection of the contractor/manager. Negotiation would be on the basis of the management fee together with the contractor/manager's estimate of site management requirements.

If the contractor/manager is to be selected in competition, tender documentation should be produced inviting the submission of proposals. This documentation will advise the tenderers of everything known about the project which, given the early stage at which contractor/managers are often appointed, may not be very much. It will, however, generally include:

- general arrangement drawings
- known specification information
- the expected contract value
- details of any key dates
- details of the contract document.

The contractor/manager's submission would normally include the following:

- the management fee
- an estimate (or tender if required) of the site management and preliminary costs
- any comments on the estimated cost of the work given the other information available
- a method statement giving outline proposals for carrying out the work, including a draft list of work packages (sub-contracts)

- a draft contract programme
- details of the proposed management team
- a proposed typical sub-contract form.

Clearly selection will be made largely on the credibility of the contractor/manager to provide the building on time and within budget as a difference of, say, 0.5% in the fee may not be significant in the context of some of the problems which could occur on a major development. With this in mind, the contractors/managers will be anxious to include in their submissions details of their past projects.

Some or all of the contractors/managers are generally invited to attend an interview with the employer and the design team principals, after which an appointment is made. This is usually on the basis of a Letter of Intent, which may be followed up by a Pre-Construction Agreement. The purpose of the latter is to define the responsibilities of the parties and the formula for reimbursement (usually a time charge for staff and actual expenditure) in the event that the project is aborted before commencement on site.

Contract conditions

All the management contracting and construction management firms and many clients, architects and quantity surveyors have for some years now had their own forms of contract. However, the JCT publishes standard forms for construction management and management contracting. These are the Construction Management Contract and the Management Building Contract respectively. Both contracts comprise a series of sub-agreements.

The Construction Management Contract comprises:

- Construction Management Agreement — CM/A
- Construction Management Tender — CM
- Construction Management Trade Contract — CM/TC
- Construction Management Trade Contractor Collateral Warranty for a Funder — CMWa/F
- Construction Management Trade Contractor Collateral Warranty for a Purchaser or Tenant — CMWa/P&T
- Construction Management Guide — CM/G

The Management Building Contract comprises:

- Management Building Contract — MC
- Management Works Contract Tender & Agreement — MCWK
- Management Works Contract Conditions — MCWK/C
- Management Works Contractor/Employer Agreement — MCWK/E

management contract

Contract administration

In management contracting and construction management it is often the case that the whole of the management team (i.e. the design team, the quantity surveyor and the contractor/manager) is based or is at least represented at senior level on site. This facilitates day-to-day communication and problem solving, consistent with the objectives of fast and flexible building. Most employers will also find it necessary to designate or appoint their own project manager, who becomes the single point of contact for the construction team.

One of the management team's first tasks is to draw up a detailed programme and procurement schedule. These show the dates by which design information is required by the quantity surveyor for the production of tender documents for each package. A list of possible works contractors is also compiled. These companies are then pre-qualified to ensure they are suitable, interested, financially able etc. and a number, usually four to six, are shortlisted to go on the package tender list. Tenders are invited by the contractor/manager and will include full preliminaries, a works-contract programme and other details such as safety and quality control procedures. Some contractors/managers also hold mid-tender interviews to make sure the tenderers understand their part in the whole project. Any significant points arising at these meetings are, of course, circulated to all tenderers.

When the tenders are received, they are very quickly reviewed by all members of the team and one or more of the tenderers may be called to a post-tender interview, especially if any queries arise. A team recommendation is then made to the employer and the package contractor is appointed.

Drawings and architect's/contract administrator's instructions are issued in the normal way. Contractors/manager's instructions are issued to the package contractors within the authority of the architect's instructions and these form the basis of the package final account. Valuations are also carried out in the normal way with package contractors' applications being checked and audited by the quantity surveyor.

Periodically reports are made by the team to the employer, reporting on progress, on any problems being encountered or foreseen, any corrective action being taken and, of course, on the anticipated final account.

Professional advisers

Management or construction management contracts may include or exclude design work but the employer will need independent professional advice, whether or not the design is included. In the case of the employer's architect and engineers, the extent of their involvement will, of course, vary according to where the design is placed. The quantity surveyor's role is very similar to normal building contracts, but the emphasis is more on cost control and auditing of expenditure by the contractor/ manager.

Advantages and disadvantages

Advantages

Among the advantages claimed for management contracting and construction management are the following:

- *Harmony, not confrontation* Traditional contracting has been described by some as adversarial. In management contracting and construction management, however, the sole objective of all parties is to produce the required building on time and on budget and the contractor/manager is paid a fee for this service in a way similar to the other members of the client's team. The process should, therefore, be more harmonious.

- *Earlier start on site* Management contracting and construction management contracts allow the project to be on site earlier than by other methods. Figure 11.2 shows that because sub-contract tenders are invited and let whilst design is still progressing, there can be more time for each of these procedures to take place whilst allowing a completion date which is earlier than by a more traditional method.
- *Design flexibility* Management contracting and construction management contracts allow the employer and designers total flexibility to develop the scheme, which enables each and every proposed design to be assessed in terms of the three factors affecting it, i.e. time, cost and quality. It is usual on a large project for the design team to be based on site.
- *Buildability* Early involvement of contractors/managers means that they can advise on the suitability of proposed materials and methods with regard to time, market availability etc.
- *Early completion* Whilst the contract states a completion date, the client can, by agreement, require the works to be completed earlier, although this may attract payments for acceleration of the various work packages affected.
- *Less pricing risk* As packages are tendered closer to the time of the work being executed on site, tenderers do not have to include a premium for pricing risk.

Disadvantages

Amongst the disadvantages are the following:

- *Possible escalation of cost* Some people express concern over cost because the contractor can be appointed whilst the design is still conceptual, but in fact almost all work packages, and many of the preliminary items, should be let following competitive tenders.
- *Inflated costs* There is a tendency for contractors/managers to wish to obtain tenders only from established sub-contractors with a reliable record. This is good for ensuring satisfactory performance but if cheapness is the most important consideration to the client, regardless of standard of workmanship and finishing times, the lowest cost may not be achieved.
- *Lack of contract sum* Whilst the estimate of prime cost is a carefully prepared and agreed budget, which is constantly refined and in which confidence will

grow as packages are let, it is not a commitment and some clients are unhappy about entering into a contract which does not have a contract sum or finite commitment.

- *Increased risk* In order to secure the contractors/manager's allegiance it is usual that he carries very little contractual risk, i.e. greater risk is carried by the client. For example, in the event of a default by a sub-contractor, the client will usually be responsible for all the effects of this on the cost and time of the project which cannot be reclaimed from the sub-contractor.
- *Duplication* There is a possibility of duplication of some preliminary items in cases where sub-contractors are each made responsible for an item which would otherwise have been provided by the main contractor. Scaffolding is an example.

Construction management

As noted earlier in this chapter, there is only one significant difference between management contracting and construction management – that of the parties to the package contracts being the employer (instead of the management contractor) and the package contractor. This was illustrated in Figure 11.1. None of the package contracts are entered into by the construction manager – all contracts are with the employer. This leads to some additional advantages, among which are the following:

- *Increased commitment* Whilst the construction manager employed will very likely be a company or partnership, it is the individual manager responsible who will affect the success of the project. He, not being a party to the package contracts, has virtually zero risk and therefore his sole allegiance is to the client and to the project.
- *Failure of construction manager* If, despite all the interviewing and checking, the construction manager fails to perform, a replacement can be put in place with comparative ease – certainly more easily than replacing a management contractor party to numerous package contracts.

These additional advantages have led to many employers preferring construction management contracts to management contracts. There are also some additional disadvantages, among which are the following:

- *No sanctions* Since there is not usually a liquidated damages provision within the construction manager's contract of engagement, depending on the precise terms of engagement, the client may have no real sanction in the event of non-performance of the manager, short of replacing him.

- *Additional involvement* Since the employer is a party to a large number of direct contracts he is responsible, with the construction manager's assistance, for the risks associated with this and the resolution of any disputes. For this reason this method of procurement is probably only appropriate for clients who regularly commission building work, have a measure of in-house expertise and wish to be involved in the detailed day-to-day progress of the works. However, since the employer should have a full complement of professional advisers in his team, this should not be a problem.

Use

Management or construction management contracts will very likely be appropriate in the following situations:

- where it is necessary to start work on site before the design is fully developed, usually because speed is the main priority
- where employers need maximum flexibility to make changes to the building or to engage their own direct contractors during the course of construction
- where time is of the essence in the project and is of a higher priority than cost
- on large and complex projects where the enhanced team effort pays dividends and the extra cost is comparatively insignificant in relation to the total project cost – on small projects these contracts can be very expensive.

Programme

As noted above, management and construction management contracts are often used when timely completion is more important then the lowest price.

Figure 11.2 shows a diagrammatic representation of the different sequence of events between single-stage selective tendering contracts and management and construction management contracts. The vertical divisions of the matrix are not intended to indicate any particular time interval – they are provided purely to assist visual comparison of the alternatives. Neither is the length of the activity bars necessarily significant, except in allowing a comparison to be drawn between the two systems. In practice, the duration of any activity must be agreed between the participants so as to achieve the employer's expectations. Nevertheless, Figure 11.2 indicates that a considerable time saving can be achieved using a management or construction management contract – although this may not be the main driving force behind the selection of this method of procurement.

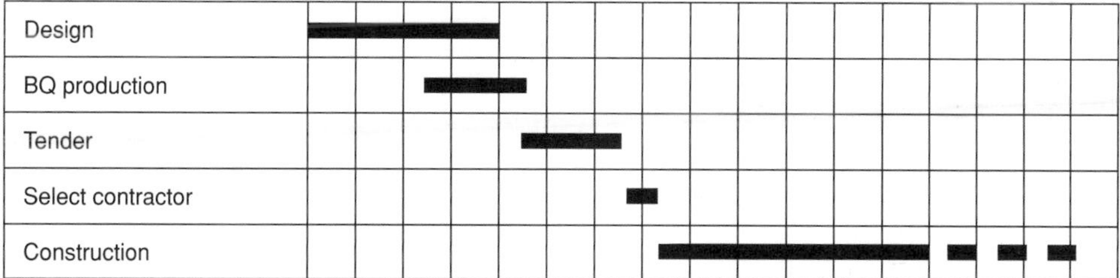

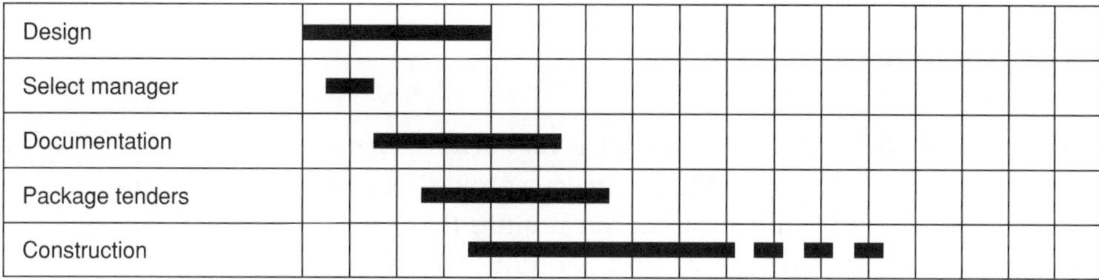

Figure 11.2 Comparative sequence of events.

Chapter 12
Design and Build Contracts

A design and build contract is a contractual arrangement in which the contractor undertakes both to design and to construct a project for a single contract sum. At the extreme, design and build projects can require that the contractor purchase land, obtaining planning permissions and consents, finance, design, procurement and construction. These contracts are known as 'turnkey' contracts and derive their name from an employer wanting to have little more involvement than simply turning the key to begin the use of a completed project. Large infrastructure projects such as power stations are often procured this way.

Design and build has emerged to be the most frequently used procurement method today. A recent industry survey for the Royal Institution of Chartered Surveyors, conducted by our practice, established that approximately 42% of the total value of the projects undertaken were procured this way.

Contractual arrangements can vary considerably. For example, during the 1970s and 80s when design and build was in its infancy, the contractor would have undertaken the complete design and construction, from inception to completion, using in-house designers and construction staff. More recently, employers have instructed independent design teams to take the brief, prepare preliminary designs, cost the project, prepare fully detailed designs and employer's requirements before entering into a contractual arrangement whereby the contractor assumes full responsibility for the complete design and construction of the works. At the point when the design is handed over to the contractor it can be anywhere between zero and 100% complete, such is the flexibility of design and build. In some instances, the independent design consultants employed by the employer may switch allegiances at a convenient time (in most cases when the design is complete) to be employed by the contractor. This contract arrangement is called novation.

Design and build contracts can be negotiated but are predominantly subject to competitive tendering as employers usually see a fixed price commitment from the contractor as one of the main advantages of the system.

'Turnkey Contract'

The Contract

There are a number of standard forms of construction contracts which cater for design and build procurement but the most widely used is that of the JCT, namely the DB Contract. However, the JCT also produces the Major Project Construction Contract (MP) which is equally suited to design and build procurement. The JCT's main traditional construction contract, the SBC, also provides for an element of contractor design under a contractor's designed portion (CDP) supplement. The CDP essentially incorporates the design provisions of the DB contract.

The DB contract embodies some very different concepts to those of the traditional JCT contracts for use with consultant design. The key agreements are set out in the first three recitals in the Articles of Agreement, as follows:

- **First Recital** This identifies the works, their location and the fact that the employer has supplied to the contractor documents identifying his requirements. These are referred to as 'the Employer's Requirements'.
- **Second Recital** This confirms that the contractor has submitted proposals for carrying out the design and construction of the project referred to in the first recital. These are referred to as 'the Contractor's Proposals'. The recital states that they include a statement of the sum required for carrying out the works and an analysis of that sum. This is referred to as 'the Contract Sum Analysis'.
- **Third Recital** This records that the employer has examined the Contractor's Proposals and Contract Sum Analysis and is satisfied that they appear to meet the Employer's Requirements.

The contractor's obligation is to design and complete the works in accordance with the details contained in the Employer's Requirements and the Contractor's Proposals. The Contractor's Proposals should expand upon the details contained in the Employer's Requirements, which can vary from a schedule of accommodation requirements to a fully designed scheme.

It is important to note that, if there is a discrepancy between the Employer's Requirements and the Contractor's Proposals, the contract dictates that the Contractor's Proposals take precedence. The contract also covers discrepancies within the individual documents – these are usually resolved in favour of the 'innocent' party.

Under the contract, the contractor is responsible for the design and bears the same professional liability as a consultant designer. The contractor is therefore bound to exercise the reasonable care and skill expected of a competent designer. The contractor is also responsible for full compliance with statutory requirements.

The contract does not provide for the employer to appoint an architect or quantity surveyor as does traditional procurement. There is instead an Employer's Agent, who acts on behalf of the employer and receives or issues applications, consents, instructions, notices, requests or statements, in accordance with the conditions.

The contract includes two options for dealing with payments to the contractor and the parties should select the preferred method, and provide the appropriate information, before signing the contract. The options are:

- Stage payments (Alternative A), on application of the contractor in accordance with the schedule of payments included in the contract particulars.
- Periodic payments (Alternative B), on application by the contractor in accordance with the period set out in the contract particulars.

The stages, under Alternative A, might include an initial payment upon receipt of planning consent (if that has been the contractor's responsibility), or commencement of work on site and then at defined stages of the project such as completion of the foundations, frame etc. Under Alternative B, the first periodic payment would have to include the contractor's costs in the early stages of the process and then continue at intervals not exceeding one month based upon valuations of the work completed.

Where to use design and build (and when not to do so)

The manner in which design and build contracts are established can profoundly affect the quality of the final product. The recommendation to use the system must therefore be based not only on the type of building required but also upon the client's expectations with regard to programme, cost in construction, cost in use, level of specification and quality of design.

Design and build contracts are certainly useful for:

- Standard building types, especially for industrial or warehousing use, and particularly where early return on capital investment outweighs considerations of design excellence.
- Buildings using proprietary systems where the manufacturer of the system might well become the main contractor, for example, repetitive housing or low cost hotels based on the assembly of factory made pods.
- Building types in which some contractors have become specialist – for example, highly serviced health care or laboratory buildings in which complexity justifies factory production.

Frequently systems of construction apply to the basic structure only, so that the choice of layout, finishes and external works can be designed to suit the client's needs without financial penalty. There may well be economic advantages to the client where a contractor's proprietary system can be used without compromising the brief.

In the public sector, where the Private Finance Initiative (PFI) is used as a means of procuring a building or project, bids are usually submitted by consortia, often led by the contractor, as the member of the team most likely to have the resources to fund the extensive pre-contract and tender work. Design, build, finance and operate (DBFO) schemes work in a similar way. The risk involved in speculative work on these projects is considerable and can affect all members of the project team – although the contractor usually takes the largest share, and receives the greatest part of the reward if the bid is successful. In such circumstances the contractor will expect to enter into a design and build contract with a corporate body especially established to procure the project. The question then arises as to the nature of the relationship between the design team and the contractor, a subject considered below.

There are, however, other types of project for which design and build is a less suitable form of procurement:

- where architectural quality is of overriding importance (the client may even wish to instigate an architectural competition)
- where clients require a building tailored to their special requirements
- where there are complex planning or environmental issues
- where complex refurbishment work (particularly of historic buildings) is required, as frequent or unexpected variations often arise
- where the brief cannot be defined, or the building function is of great complexity, so that a protracted period of research and investigation is necessary at the outset and might continue once work has commenced.

Whilst the issues above might be construed as a disadvantage to design and build as a method of procurement, there are nevertheless means of resolving most of them. An employer may use a design team in the traditional sense to develop the brief before handing a completed design over to the contractor under novation. The use of design and build contracts is principally a risk driven process where the contractor assumes the legal risk for both design and construction.

Managing the design process

The key to any design and build contract is the brief or 'Employer's Requirements'. This may be in the form of an outline specification, a performance specification, drawings in varying degrees of detail or drawings and an outline specification in combination.

The employer often appoints a design team to assist in compiling this brief and preparing preliminary designs and costings before seeking a contractor by competition or negotiation to proceed on a design and build basis. The costing element of this exercise is important as it is vital that the contractors tender on a basis that the client knows is within the budget.

Design and build is an expensive procurement route for contractors to tender in competition, as each has to work out a detailed design to suit the brief and a price to go with it. Clearly this can be an uneconomic use of resources, which will reflect on the ultimate cost to the employer – especially if too much detail is required as part of the competition process. Before contract, therefore, contractors should go no further than an outline scheme design based upon a specific and concise brief. Detail design, and fully detailed specification come later.

Once the contractor has been appointed and has assumed responsibility for the detailed design, the decision must be made as to the future role of the employer's consultants. When the contractor uses an in-house design team to develop the details for construction purposes, the original designers and quantity surveyor might be retained by the employer to monitor standards and supervise payments. Alternatively, the original designers might be 'novated' to the contractor, leaving the quantity surveyor to give cost advice to the client and/or act as the Employer's Agent.

Novation

Novation is a legal term used to describe a procedure whereby one contract is substituted for another – commonly where a contract between two parties, A and B, is replaced by a contract between parties A and C. Novation is widely used on design and build contracts, when the contract between a consultant and the employer is replaced by a contract between the consultant and the contractor. However, as noted below, the terms and conditions of the substituted contract are, of necessity, different to some extent.

Novation will normally take place as the contract is let. The Employer's Requirements, issued as part of the tender documents, should clearly state that the consultant's appointment is to be novated, and include the precise terms and conditions of the consultant's original appointment. Since the consultant's original appointment will frequently include the provision of services which have been completed, or which are not relevant to the new appointment, the terms and conditions of the new contract are bound to be different. For example, feasibility studies and sketch proposals, forming part of the original appointment, will already have been completed and will not be required as part of the new contract, and the contractor will not necessarily require a post-contract management service. A schedule identifying the services which are to be provided under the new appointment should be drawn up and agreed by the consultant, employer and contractor.

Difficulties often arise when employers fail to realise that novated consultants no longer have duties to perform for them. In practical terms, the consultants are no longer employed by the employer and are therefore no longer able to represent the employer's best interests, monitor the quality of construction or deal with payment. These duties must be left to others, such as the Employer's Agent. With due advance planning, however, most potential issues can be anticipated. Depending on the extent of the design responsibility included in the contract and the contractor's own professional indemnity insurance cover, the consultants may be required to enter into collateral warranty or third party rights agreements protecting the employer against damages arising out of design based failures. These indemnities might only be called upon in the event that the contractor becomes insolvent since, following novation, the employer has a direct contractual link with the contractor for guaranteeing adequacy of the design.

Evaluation of submissions

When design and build is used in its pure form, whereby the contractor undertakes to carry out all of the design, once the design and build tenders are returned the evaluation of the contactor's proposals is a complex task. Each contractor will almost certainly have interpreted the brief in a different way making direct comparison difficult. However, in most cases, the submissions can be compared on a common or equivalent basis by adjusting those items which are provided for in one submission but left out of others and comparing overall value for money. In some cases, this is very simple, for example different floor coverings can easily be evaluated or costed and therefore compared and contrasted.

Design qualities may be a significant factor and in the case of a design competition are obviously paramount. In this scenario, evaluation is very subjective and cannot be expressed in purely monetary terms. Sound professional advice is necessary from independent architects, engineers and quantity surveyors.

After selection there will be a period during which the contractor will be refining the details of the design, preparing documentation for obtaining sub-contract tenders and mobilising site resources. Frequently, the period of development of the design details will bring to light matters which can be incorporated in the building without penalty and for the benefit of the employer. All these matters must be taken into account when finalising the formal contract documents in support of the agreed contract price.

Post-contract administration

Although the total responsibility for carrying out the design and construction work rests with the contractor, the employer is normally well advised to monitor the contractor's performance to ensure that the specification is adhered to, as set out in the contract documents, and that the workmanship is to the requisite standard. If the original consultants have not had their contracts novated to the contractor they will be able to do this; if novation has taken place the employer must rely on others.

Although there is no provision in the contract for this role (in terms of an architect or a clerk of works), the Employer's Agent has reasonable access to the works and the employer has powers to order the opening of covered work for testing.

Financial administration

Care must be taken to ensure that the contractor is paid the correct amount for interim and final payments. Again the employer will be well advised to monitor the contractor's performance, for under this contract all the financial calculations are undertaken by the contractor. The contract does not make provision for an employer's quantity surveyor, although the employer still has the right to challenge the contractor's calculations.

The contractor is charged with making applications for payment, using either the stage payment or periodic payment system. Valuations made by the contractor will include the valuation of variations which must be priced according to the valuation rules laid down in the contract. This, as with other JCT contracts, is a unilateral activity but in this instance, of course, it is the contractor who does the calculations and not the employer's quantity surveyor.

It is usual for the employer to appoint a quantity surveyor as financial administrator, and preferably to retain the services of the quantity surveyor who has been involved in the preparation of the brief, or Employer's Requirements. It is not uncommon for this appointment to coincide with that of the Employer's Agent.

Programme

The main driving force behind the use of design and build contracts is probably the desire to have a single point of responsibility, closely followed by the hope of a reduced risk of costs rising. However, a reduced timescale for the project is also a possibility.

Figure 12.1 shows a diagrammatic representation of the different sequence of events between single-stage competitive tendering and design and build contracts. The vertical divisions of the matrix are not intended to indicate any particular time interval – they are provided purely to assist visual comparison of the alternatives. Neither is the length of the activity bars necessarily significant, except in allowing a comparison to be drawn between the two systems. In practice the duration of any activity must be agreed between the participants so as to achieve the employer's expectations.

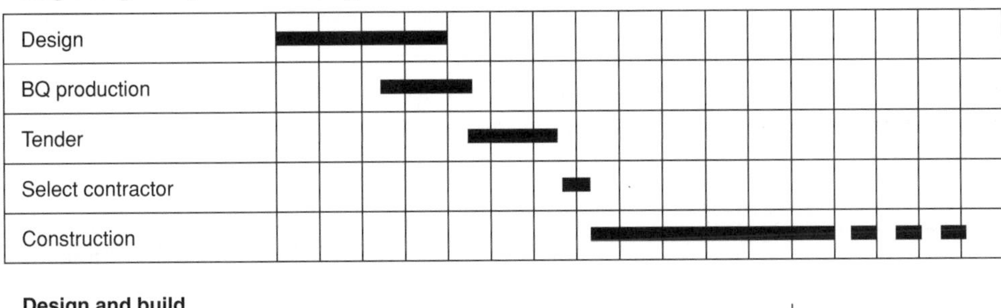

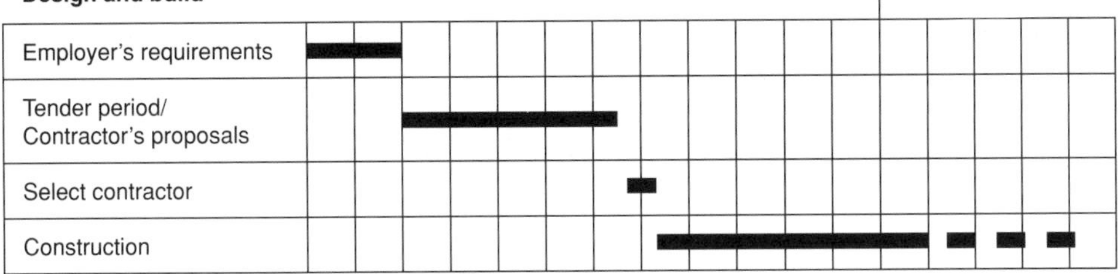

Figure 12.1 Differing processes compared.

In Figure 12.1 the overall project timescale is shown as the same for both systems but, as noted above, this is not intended to be significant. Depending on the size and complexity of the project, and the client's priorities, the overall timescales can be varied. However, many design and build specialists claim that programmes can be shortened because the process is closely co-ordinated within their single responsibilities and because of the shortened learning curve – team members are used to working together.

Advantages and disadvantages

Advantages

- Early certainty of overall contract price is obtained, as long as the JCT form of contract is used.
- Responsibility for the design, construction and the required performance of the building lies entirely with one party, the contractor.
- Design and build imposes a discipline, not least on the employer, to define the brief fully at an early stage. Given this, the advantages of overlapping design with construction can lead to a shorter project duration.
- Given the higher degree of co-ordination at an early stage, through the single point of responsibility, variations during construction tend to be fewer and the risk of post-contract price escalation reduced.

Disadvantages

- Design and build, by its very nature, is a rigid system which does not allow the employer the benefit of developing requirements and ideas, despite the facility for accommodating variations.
- Where several tenders are invited comparison can be difficult as the end product in each case is different, and the final choice will be influenced by subjective judgment.
- Any variations required by the employer after signing the contract can prove expensive and difficult to evaluate.
- There is always a risk with regard to the quality of work. If the original brief is not precise and the specification offered by the contractor equally vague there is a temptation for the contractor to reduce standards.
- If the design team's contract is novated to the contractor a conflict of interest can arise.

- Following novation of the design team's contract to the contractor, employers have no direct input for checking or improving quality unless they appoint other consultants.
- Most design and build contracts are qualified in some respects (e.g. by ground conditions or the inclusion of provisional sums) which to a certain extent negates the employer's ideal of an early known financial commitment.

Chapter 13
Continuity Contracts

In Chapter 4 several aspects of the economic use of resources were considered. Among these was the subject of continuity. Continuity has two characteristics which are of interest in this context: first, economies of scale and, second, the effect on the learning curve. Both characteristics have potential favourable cost implications – the first directly in reduced material costs derived from greater buying power, and the second indirectly through a reduction in timescales and thereby reduced labour costs.

In previous chapters, various methods of procurement and contractual arrangements have been discussed in relation to single projects or construction sites. In circumstances where continuity is possible advantage can be gained by adopting a different approach – by linking different contracts, or work packages, together.

The greatest scope for applying such concepts is obviously with clients or agencies that have a regular or continuing building programme. This can be for one or more new buildings, or in maintenance contracts which require the same repetitive type of work to be undertaken on a number of buildings and/or for a certain period of time.

In this chapter, the purpose and use of three types of continuity arrangements are compared and their operation is described to illustrate how benefits can be gained. The three types are:

- serial contracting
- continuation contracts
- term contracts.

Serial contracting

In many programmes of building work, such as providing several supermarkets for a supermarket chain or refurbishing a high street shopping chain for a retailer, there is an element of continuity. The object of any tendering procedure should be to ensure that building resources, including associated professional resources, are used as economically as possible. Where there is continuity these resources may be used more economically and efficiently if all the work is carried out by the same contractor, rather than by having a different one for each site or operation. It is important, therefore, to identify where the continuity lies and to evaluate the benefits that will accrue if one contractor undertakes the work instead of several.

As well as the economies of scale and reduced head office overheads because of the repetitive nature of the work, contractors with the knowledge of security of future work are likely to work for reduced margins since wider industry market risk is reduced. In much the same way as a client benefits financially through continuity, the contractor can also secure financial benefits through greater efficiency in the control of the supply chain.

However, it is not enough merely to establish that there is a saving in resources. As the clients provide the continuity by virtue of the programme, they should gain more than just financially by securing consistent quality and good working relationships. It will be appreciated that in any situation where one contractor is to do a series of jobs for a client, a good relationship becomes particularly important.

Serial contracting has been broadly defined as an arrangement whereby a series of contracts is let to a single contractor, but further definition is needed to distinguish it from other types of continuity arrangement. In a serial contract the approximate number and size of projects is known when the offers are obtained and the individual projects tend to be of the same order of size. The serial tender is a standing offer to carry out a series of projects on the basis of pricing information contained in competitive tendering documents. These may consist of bills of quantities, specifically designed as master bills to cover the likely items in the particular projects envisaged.

The series will usually consist of a minimum of three projects; the number will normally be known at the time of the tender, although further projects may be added by agreement at a later stage. Final designs will not necessarily be ready for all of them at tendering stage but clients and their architects will have a good idea of the typical requirements, and these will be reflected in the tendering documents prepared by the quantity surveyor.

Exact quantification will be achieved as the design for each project in the series is finalised. Normally a separate contract based on the original standing offer will then be negotiated and agreed for each project, based on standard items extracted from the master bills. All new or 'rogue' items will be negotiated for each contract as they occur.

Purpose and use

Serial contracts are ideally suited to a programme of work where the approximate number and size of projects to be contracted is known at the time of going to tender. For example, in refurbishing a series of high street shops for one client it will normally be the quantity of work which varies rather than the quality or specification. Similarly, although supermarkets for a supermarket chain may vary in size, in a new building programme even the largest is likely to be less than three times the size of the smallest. Moreover, the sizes of most of the projects are likely to be far closer to each other than this. Therefore the same type of contractor management is likely to be needed for them all. It is probably this latter point which should be regarded as one of the governing criteria.

Consideration should be given to the choice of this method of contracting in times of high inflation, or when there is a building boom, when it may be difficult to secure a firm price commitment over a lengthy period of time.

Operation

On serial contracts the tendering procedure will normally be on the basis of a master bill which will comprise most of the items envisaged in the various projects in the series. Individual projects will be controlled by having separate contracts negotiated on the basis of the master bill, after which work will proceed very much as for a normal contract.

It is desirable to take great care in the selection of the contractors for the tender list before formal tendering takes place. A mistake made when dealing with a series is likely to have more serious results than in the case of a one-off contract, so it is much more important to establish the financial and physical resources of the contractor. A bankruptcy, covering as it would many contracts rather than one, would be disastrous. It could be advantageous to make provision within the contract for the client to be able to withdraw from a contract if performance is found to be below par. A further point to note is that small errors of documentation or detail, which may not amount to much on one contract, are clearly multiplied in a serial contract.

Comparing the contractor's capacity with the likely programme of work also requires great care. Although the total series may be large, the individual contracts may be relatively small. It is possible, therefore, that a contractor who might not be able to cope with such a large contract on one site would be able to cope with a number of smaller projects spread over several sites and, possibly, over a longer time. It is important, however, that the contractor's physical resources should not be stretched to the point where it is difficult for all the work to be completed. Any unprogrammed overlaps in projects could present a contractor with resourcing problems. If significant changes in the overall programme occur it may be worth considering whether or not to retender.

It is likely that there will be a stage before formal tendering where interviews are held with possible tenderers, references sought, and a short-list drawn up on the basis of the information obtained.

The result of competition should be that the saving in building resources arising from the continuity is passed to the client. It is emphasised, however, that this is only possible where the extent of the work to be covered by the serial contract is known sufficiently for reasonable estimates of prices to be given in the original standing

'... small errors of documentation ...'

offer. It should be noted that the savings derived from serial contracting are evidenced in all projects – the savings are contained in the rates in the master bill used to price all individual contracts.

Continuation contracts

A continuation contract differs from a serial contract in that it results from an ad hoc arrangement to take advantage of an existing situation. Thus, if a contract is already going ahead on a particular phase of a business park site, the opportunity may arise at that point to commence a further phase with similar requirements. In this case there may have been no standing offer to do more work, the original tendering documents having been conceived only for one particular phase. However, those documents can provide a good basis for a continuation contract. Alternatively, it is possible to make provision for continuation contracts in the tendering documents for the original project, and this is sometimes done. There is, however, no contractual commitment, and in the event a continuation contract may not arise.

In either case the contractor has to price the tender documents for the original contract on the basis of getting that particular contract only. The continuation contract, if and when it arises, is then dealt with separately – and it is only at this stage that savings arise.

Purpose and use

As with serial contracts, continuation contracts are best suited to situations where the scope, style and nature of the work are very similar. The more variations there are between the projects, the more variations there will be in the contract documentation. This will result in an increased number of purely negotiated costs (rather than adjustments of competitive prices) and could also lead to organisational problems on site.

The purpose of a continuation contract should be to seek a financial advantage for the employer. Continuation contracts offer the employer the following potential benefits:

- a faster start on site, resulting from the shorter pre-contract period
- a competitive basis for pricing, resulting from the initial tender
- an experienced contractor, who has identified (and solved) the construction problems of the previous contract (this represents perhaps both a cost and a time benefit)
- re-use of the contractor's site organisation and management team.

Whilst circumstances do arise where the use of continuation contracts might be appropriate, such as phases of a business park, the practicalities of a situation often preclude this. For example, whilst the style of the work is very similar the projects would not necessarily be of a comparable size. Additionally uncertainties such as planning approvals, existing leases/tenancies and enabling/preparatory works often make site acquisition and site handover ready for construction a precarious job and all create obstacles to continuity.

However, provided there is a degree of parity between the projects a continuation contract may be appropriate where serial contracting is not.

Operation

In continuation contracts, the tendering documents will obviously be closely related to the documentation for the original project on which the continuation is to be based. The contract sum is negotiated on the basis of the original contract but with two substantial adjustments.

First, the contractor may require increases in the cost of labour and materials to be taken into account, to allow for the time difference between the two tenders. The only exception to this might be where the original contract was let on a fluctuations basis – the continuation contract could then, if necessary, be from that same base and the increases would be covered entirely by the fluctuations clause. The same would apply, of course, to any decreases.

Second, employers will wish the benefits of the continuity they have provided to be passed to them in the continuation contract. These will be of two kinds:

- Productivity in the construction industry generally has increased year by year and, therefore, if the continuation contract is timed, say, nine months to a year later than the original contract, there should be a saving resulting from increased productivity.
- There is a further saving arising from the economies inherent in the same contractor doing similar work for the same architect and client. This element varies from one contract to another, but may be quite substantial where similar houses or flats are being built on the second site.

The matter of measurement of productivity is difficult, and requires much research. A lot will depend on the information available to the quantity surveyor, and that person's skill in analysing and assessing elements which affect productivity and in negotiating with the contractor on this basis.

One of the prerequisites for the successful negotiation of a continuation contract is parity of information between client and contractor. This involves the contractor disclosing the details of how the tender for the original contract was built up. If the contractor is not prepared to do this then it may not be prudent to proceed on this basis.

Once the continuation contract has been negotiated, a figure agreed and a contract signed the project will proceed as any other. There is one factor, however, that should be considered, particularly where it is envisaged that several continuation contracts may arise from the original. The contractor should be given an incentive to make cost reductions. Where continuation contracts are being considered, the contractor might well be given the full benefit of any reduction in cost he makes on the first occasion, even though this involves a change in specification (but not, of course in performance), with the proviso that the full amount of the cost reduction be allowed to the client in the succeeding continuation contract. This situation arises particularly where the contractor is involved in a considerable amount of design work, and where the detailed specification is very much related to his production requirements.

Term contracts

Term contracts differ from both serial and continuation contracts in that they envisage a contractor doing certain work for a period of time or term. In this situation the contractor agrees a contract to do all work that he is asked to do within a certain framework and during a given period. Term contracts are generally set to run for a period of 12 months, but a longer period is likely to result in better tenders. Experience suggests that a term of up to three years can generally be expected to promote confidence within the period of the contract. However, the volatile nature of building costs is such that provision should be made for the employer to test the market at shorter intervals. Thus a period of two years may, on balance, be regarded as the optimum for such contracts.

Purpose and use

Term contracts are well suited for use with the management of large and continuing programmes of day-to-day reactive building maintenance. The management and control of such programmes can be an extremely complex task and in view of their potential for flexibility term contracts are perhaps best suited to the problem.

Care should be taken both in the compilation of the tendering documents and in their evaluation. Generally speaking the main tendering document will be a schedule of rates. Additionally, it can be helpful to prepare a bill of provisional quantities, based on previous years' workflow, to give an indication of the likely volume of work anticipated on specific trades over the term of the contract. As always, it is advantageous if contractors of like characteristics and performance can be selected to submit tenders, and it will usually be an advantage if multi-trade contractors can be utilised in view of the many operational, as opposed to single-trade, jobs to be found in building maintenance.

Term contracts may also be used for painting and redecoration, roadworks and other specialist trades.

JCT Measured Term Contract

At the request of its constituent bodies the JCT produced the Standard Form of Measured Term Contract in 1989 which has recently been revised and published as the Measured Term Contract in 2006. It is specifically intended for employers who have programmes of regular maintenance and minor works, including improvements.

This form requires the employer to list the properties in the contract area and the type of work to be covered by the contract, as well as the term for which it is to run. The employer is required to estimate the total value of the contract and to state the maximum and minimum value of any one order. There is provision for priority coding of orders to deal with emergencies and specific programming requirements.

Payment is made on the basis of the National Schedule of Rates, or some alternative priced schedule, and the contractor quotes a single percentage adjustment to the base document, or perhaps different percentages for different trades. Provisions are included for fluctuations and dayworks. Measurement and valuation of individual orders can be carried out by the contract administrator or by the contractor, or can be allocated to either party according to value.

It is also possible to compile an ad hoc schedule, relative to any employer's particular range of work.

Operation

Term contracts are ideal for carrying out maintenance and repair work. In this type of work the individual project can be very small, say from £500, while the largest may be in the region of £30,000 or more. Tenders can be sought using a large number of methods but two of the commonest are as follows:

- The employer can determine the unit or operational rate for each item and the tenderers then offer to do the work on a plus, minus or zero percentage basis to reflect their bid.
- Alternatively, tenders can be sought on the basis of blank schedules which the tenderers then price.

Analysis of the tenders will usually be found to be easier if the first method is used; depending on the number of tenders received the analysis of priced schedules can be a daunting task. Experience will suggest the best method for individual circumstances.

Additionally, as part of the contract, tenderers may be required to meet specified response times for various categories of work, including emergency items. The ability to meet these response times will be an important consideration in the selection of the successful contractor(s).

Contracts can be entered into with one or more contractors and the work placed on the basis of cost, committed workload and performance. Generally speaking, the more contractors available to do the work, the smoother the operation of the maintenance programme. Orders for work are issued from time to time; the work is agreed and the contractor paid accordingly. The number of orders issued annually under a maintenance term contract can run into many thousands. To ensure the workload can be dealt with smoothly and expeditiously continual monitoring of the contractors' performance will be necessary and orders to the various contractors regulated accordingly.

Orders should be given a priority rating, which will require contractors to attend and carry out the work within a specified time. An incentive scheme will be found to be helpful to achieve this. Where the work is completed within a specified time, an attendance payment can be made. Alternatively, and perhaps more effectively, where the contractor fails to complete the job on time, the number of orders can be reduced until performance improves again. Additionally, a plus payment can be made for attendance to emergency call-outs. Emergencies which relate to health and safety matters can arise at any time and will usually require quick attendance – initially to make safe only, the permanent repair being dealt with separately.

The average cost of day-to-day maintenance items will be found to be something up to £100. Assuming that the builder deals, on average, with say, 100 orders per week, his average monthly account will be in the order of £35,000 to £40,000, a not inconsiderable cashflow. It is therefore incumbent on the employer to ensure that payment of the contractors' accounts is made expeditiously.

Receiving repairs requests, deciding on remedial action, determining priorities, allocating work, monitoring attendance and completion times and inspecting and certifying payment can be a complex exercise. The use of a sophisticated computer program, which should be based on a property register and linked with a comprehensive schedule of rates, together with an order generator and accounts for payment, will be found to be invaluable in the smooth running of a maintenance based term contract.

For an employer with a large estate and a large number of minor works of this nature to be carried out, a term contract is a method of controlling the work with a measure of accountability and a simple method of procurement for the individual projects.

Chapter 14
Partnering

Partnering is perhaps seen as being a relatively new approach to the procurement process, in which an attempt is made to improve relationships and performance for the benefit of clients and the other members of the team – what is termed, in management parlance, a 'win-win' situation. Many clients are dissatisfied with the results of individual competitively tendered projects. These contracts have often underperformed in terms of time, cost and quality and have led to adversarial attitudes from all parties.

Partnering

Since reference to partnering was made in Sir Michael Latham's report *Constructing the Team* and Sir John Egan's report *Rethinking Construction* there has been a steady line of publications advocating its use. These, however, are only manifestations of a process which has been developing for some considerable time, although the name 'partnering' may only have been applied to it recently. It is said that the oil, chemical and power generation industries in America first used the system, while in the UK the British Airports Authority's framework agreements and, more recently, the National Health Service's Procure 21 can be considered as forerunners to what is now recognised as partnering.

All these reports have thrown out the challenge to the industry – to improve relationships and to deliver measurable improvements in time, quality and cost.

A definition

The terminology encountered in this field: 'project partnering', 'strategic partnering' and 'first, second or even third generation partnering', or even 'collaborative working', gives a clue to the variety of interpretations placed on the term 'partnering' and therefore the impossibility of a finite definition.

Partnering has been described as a structured management approach which facilitates teamworking across contractual boundaries by integrating the project team and smoothing the supply chain. The three fundamental characteristics (see Figure 14.1), are:

- formalised mutual objectives (which may be binding or non-binding) of improved performance and reduced cost
- the active search for continuous measurable improvement, which is perhaps measured against industry key performance indicators (KPIs)
- an agreed common approach to problem resolution.

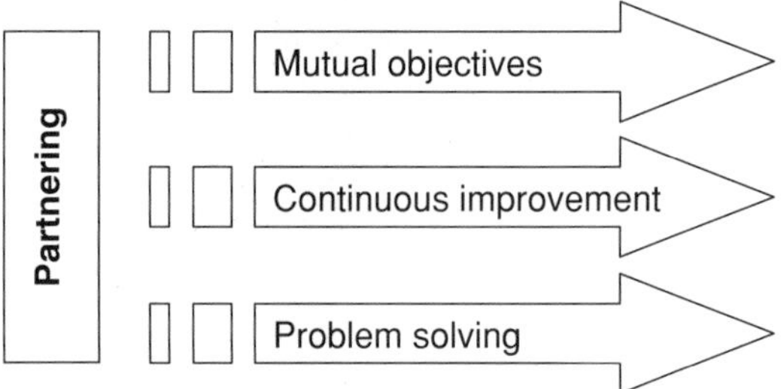

Figure 14.1 The three fundamental characteristics of partnering.

Partnering should not be confused with other good project management practice, or with longstanding relationships, serial working, negotiated contracts or preferred supplier arrangements, any of which may be present but all of which lack the structure and objective measures that support true partnering.

The essential characteristic of partnering is the commitment of all partners at all levels to make the project a success. The result is that the partnering agreement between the parties drives the relationship, rather than the contract documents. This is not to say that the contractual relationships are not important – they must be in place for the day-to-day running of the project. Standard forms of contract can be used in partnering, although it is as well to check the specific terms of any document to make sure that they do not cut across the aims of the partnering agreement.

Partnering can be applied to single jobs, often then termed 'project partnering', but the benefits of partnering tend to be cumulative, so that significantly greater benefits arise over several projects – a situation often termed 'strategic partnering'. It is also considered that the benefits to the client are further enhanced if partnering is applied throughout the supply chain, thus covering the employer, the professional advisers, the contractor and sub-contractors and the suppliers. If the client's workload is sufficiently large or geographically dispersed, there is no reason why several different partnering agreements should not be set up, running concurrently, with different groups of partners.

It should be borne in mind that, according to EU procurement regulations for public sector procurement, partnering can only be applied to projects after the award of the contract. The European Construction Institute's document *Partnering in the Public Sector: A Toolkit for the Implementation of Post-Award, Project Specific, Partnering in Construction Projects* is particularly relevant in this context.

The key to partnering is a change in mental attitude, a new approach based on honesty, co-operation and deeds (not just words). Without a fundamental change in attitude incorporating these three ingredients, no partnering agreement will work, or achieve the claimed significant improvements in performance.

When to adopt a partnering approach

Partnering is applicable to all projects in terms of supply chain management and being able to obtain a mutual benefit for both parties. However, there is perhaps more scope for partnering in circumstances where there is a high likelihood of inefficiency undermining the project. This is said to be an advantage of the JCT's non-binding framework agreement which adopts a partnering ethos where working to a rigid binding agreement might add to the project cost should the project not be sufficiently large enough to absorb the cost of a learning curve. Partnering is said to be particularly appropriate in situations where:

- the project is technically complex and difficult to specify
- the business requirements require advice upstream and downstream of the supply chain
- the client has projects of a similar nature giving scope for continuous improvements in terms of time, cost and quality
- construction conditions are uncertain and solutions are difficult to foresee giving rise to an element of joint problem solving.

The agreement

A partnering agreement may be a verbal agreement, a broad written arrangement on a single piece of paper, or a more detailed and formal document. Each situation will have different aims and requirements, which will dictate the appropriate method of setting up the partnering agreement. Care has to be taken, however, to ensure that a detailed written agreement does not lead to an adversarial attitude if things go wrong. One might also question what benefit there is in a document which comprehensively codifies the expected behaviours of the project participants because the need to do so may underscore the point that the behaviours are not those which would instinctively be followed.

The intention of a partnering agreement is to establish a framework, based on mutual objectives and agreed procedures, allowing the participants to maximise specific business goals. A partnering agreement should consider the following areas:

- The formulation of a statement outlining the philosophy of the parties – that of working in good faith.
- The setting of realistic and achievable targets, with a procedure to monitor and review these as necessary.
- The agreement of a method where open-book costs can be established and savings can be shared by all participants.
- A procedural framework, including the definition of the parties' roles, responsibilities and lines of communication.
- An agreed procedure for the resolution of disputes, without damaging the original objectives of the partnering agreement.

JCT Framework Agreement

The JCT does not specifically publish a 'partnering' contract by name but provides for two framework agreements that embrace all of its virtues. The only material difference between the two framework agreements is that one is legally binding

(FA) and the other is legally non-binding (FA/N). It is worth reflecting on the JCT's intention as to the role of the framework agreements:

> *'The aim of this framework agreement is to provide a supplemental and complementary framework of provisions designed to encourage the parties to work with each other and with all other project participants in an open, co-operative and collaborative manner and in the spirit of mutual trust and respect . . .' (Clause 4)*

Both framework agreements provide for nine core framework objectives and contain provisions advocating mutual benefit for the parties. These are:

- zero health and safety incidents
- teamworking and consideration of others
- greater predictability of out-turn cost and programme
- improvements in quality, productivity and value for money
- improvements in environmental performance and sustainability, and reductions in environmental impact
- right first time with zero defects
- the avoidance of disputes
- employer satisfaction with product and service
- enhancement of the service provider's reputation and commercial opportunities.

The legally binding framework agreement contains provisions for the settlement of disputes which is, perhaps, somewhat of a paradox in terms of the partnering ethos.

The partnering workshop

The practical details of the partnering agreement, its procedures and objectives are often worked out in a one or two day workshop, where all the participants come together at the earliest possible stage to plan the project. This is the start of the teambuilding exercise and involves the actual personnel to be involved in the project, not just the senior management or marketing staff.

The workshop is often run by professional facilitators and terminates with the drawing up of the project charter, which should be signed by all participants.

The benefits

A properly set up and effective partnering agreement should seek to achieve benefits in respect of time, cost and quality. Such benefits will be derived in a number of ways, based on the following concepts:

- **Reduced learning curve** Optimum performance and productivity can be achieved much earlier in the process due to familiarity with working practices and confidence in the expertise of other participants.

- **Reduced abortive tendering costs** Overheads involved in tendering are reduced. In traditional competitive tendering a success rate of perhaps one in five or six is not unusual; this adds a heavy burden to overhead costs.

- **Administrative efficiency** Continuity of personnel and familiar systems of management eliminate teething problems and silly mistakes.

- **Improved communications and decision procedures** Structured procedures, developed during regular contact, promote confidence and reliability.

- **Improved quality and programming** An overall strategy giving a proper balance to the client's priorities avoids the achievement of one target at the expense of others, so quality and time are not necessarily sacrificed to cost.

- **Economies of scale** Advantage can be taken of continuity and preferred supplier arrangements, so partnering exists throughout the supply chain.

- **Continuity of work and personnel** Long-term relationships can be developed, building confidence and reliability.

- **Risk** Risk identification is improved, leading to better risk allocation and management.

- **Problem solving** Problems are identified at an early stage, thus facilitating an early solution and avoiding escalation. If a formal dispute does arise, agreed procedures can be implemented quickly to achieve an early resolution.

The risks

Partnering arrangements are no panacea for the ills of the construction industry and the benefits do not come automatically. Partnering needs commitment at all levels in the organisations involved if the benefits are to be fully realised. However, the parties must be vigilant to prevent complacency setting in. Among the risks attached to partnering, like many other innovative systems, are the following:

- **Potential lack of accountability** The absence of competition means that other mechanisms are required to demonstrate the absence of collusion; quality and audit systems and the realisation of mutual benefits should satisfy both parties.

- **Unrealistic targeting** The search for constant improvement can lead to targets for cost, quality and time being either too low or too high; again the open-book approach to management should provide reassurance.

- **Commercial pressure** The desire to increase returns on the part of contractors (or their shareholders) may tempt them not to declare openly the true cost savings achieved.

- **Cost attitude** Having removed, or reduced, the adversarial approach in favour of teamwork, care must be taken to avoid slackness and the taking of short cuts. As noted above, the ordinary contract rights and responsibilities must continue to be observed.

- **Change of personnel** It is impossible to prevent some turnover of staff and there is a danger that new members of the team will not understand the concept, or not be as committed to the project as those they replace.

- **Benefits slow to materialise** The full effect of the benefits may not appear immediately; indeed some will not materialise unless a number of projects are undertaken. Careful analysis of the potential benefits, accurate targeting and proper evaluation of the feedback are required to avoid disillusionment should results be slow to materialise.

- **Programme amendments** When the employer's requirements or priorities change, or the programme of projects is varied, the mutual goals may no longer be achievable. It is important to analyse fully the effect of any variations in the brief on all members of the partnering arrangement.

- **Dissolution on failure** The termination of any partnering arrangement, for whatever reason, may lead to financial problems when unravelling complex relationships.

Future of partnering

For partnering to be successful, it requires a more flexible attitude and approach than has traditionally been the case in the industry, and an acceptance that organisations and individuals are entitled to receive a reasonable return for their efforts. It also requires people to accept that the cheapest price does not necessarily provide best value for money. A solid base of traditional training and good professional and commercial practice needs to be extended within the industry to allow the more lateral approach of a partnering arrangement to benefit all concerned.

The essence of partnering is in the attitude of the participants, and in their ability to achieve the optimum overall solution given a particular set of requirements and circumstances, while ensuring that all participants benefit. This will be achieved by matching one's deeds to the words – all too often, people sign up to partnering agreements then act as if they have not, only to express dismay that the partnering agreement has not spared them failure.

Part III

Preparing for and Inviting Tenders

Chapter 15
Procedure from Brief to Tender

Initial brief

The initial brief from the employer may be little more than a statement of intent, as mentioned in Chapter 2. However, for the project to be successful, the employer's brief must be developed in detail so that due consideration can be given to each of its aspects. The design brief and the design process develop in an iterative manner, with progress on one aspect creating a need for further thought on and consideration of other matters. The more complicated the building, the more important it is that the brief is developed in a systematic way and as early as possible. Full development, certainly on major schemes, is a team affair requiring the preparation of feasibility studies, cost reports and consultations with planning and other relevant authorities.

Developing the brief

Numerous factors need to be taken into account when developing the brief. Some of the main ones are:

- **Employer attitude** Establish whether the employer requires adherence to strict corporate standards.
- **Environment** Determine the type and quality of the internal and external environment desired.
- **Operational factors** Determine the activities the building is to accommodate and the other practical factors that will govern its layout and the relationship of the various elements within it. For example, assess the need to isolate certain items of plant and machinery in order to create acceptable noise and other environmental conditions elsewhere in the building. Also determine the extent of flexibility required in the designed space to provide for future adaptability.

- **Site** Identify ownership and any operational hazards. Evaluate access limitations, site topography and soil suitability (using soil tests, trial and bore holes). Survey the site and/or buildings to provide critical information such as access, boundaries, watercourses, trees, existing services, dimensions, levels and condition.
- **Timescale** Assess the appropriateness of the desired commencement and completion dates in the context of the practicalities of the situation and other projects that the employer may be planning and determine a programme.
- **Finance** Establish the employer's budget and funding arrangements and whether or not the envisaged scheme is likely to be varied if additional funds become available. Determine the relationship between capital and revenue funding and any cash flow constraints.
- **Market** Examine the market potential, timing and selling strategies for speculative buildings.
- **Costs** Assess overall projected capital and running costs.
- **Development control** Determine whether any particular development control considerations affect the site. Liaise as necessary with the development control authority.

Feasibility stage

Developing the design and preliminary cost estimate will lead to a feasibility report, advising the employer whether the project is feasible functionally, technically and financially.

Sketch scheme

The design process usually commences with the sketch scheme. Initially this involves the designer putting on paper preliminary responses to the brief. The sketch scheme will demonstrate the extent of accommodation that can be achieved: for example, the number of new houses on a particular site or the number of shopping units with their associated parking requirements. When the emphasis is on design quality as it would be in a conservation area or in the case of extensions to sensitive buildings, the sketch scheme will need also to address the visual form, style, proportion and materials. Whatever the nature of the drawings or information produced at this stage, the presentation should be capable of being readily understood by those reviewing the proposals.

Unless the employer has dealt with the issue, members of the design team should now advise on development control matters. They must determine whether or not the proposed development will require development control approval. If not, it is wise to obtain written confirmation of this in case of later dispute. Assuming approval is required, the need to obtain an outline development control approval prior to detail approval should be discussed. This usually becomes relevant when a principle needs to be established, for example a major change of use, the development of a site not designated under a local authority plan, or the development of a site over and above local density levels. Unless in a sensitive area, only basic plans or Ordnance Survey map extracts are required, thereby limiting the employer's financial commitment.

Costs

At the start of the sketch scheme the quantity surveyor will probably be required to prepare an initial cost estimate; inasmuch as the sketch scheme will be short on detail yet the employer will expect a reasonably indicative estimate of cost, it will take an imaginative quantity surveyor to address the full cost consequences of outline design information. This will invariably be based on a simple cost per square metre calculation

'... an imaginative quantity surveyor ...'

and/or a comparison of outline unit costs. As the scheme develops it will be possible for the initial cost estimate to be refined on an element-by-element basis (see Chapter 16) such that the predicted cost may be amended or confirmed. This will establish whether or not the scheme is likely to be within the employer's budget. If it is not then appropriate adjustments can be made – these can be done more readily at this stage than later in the design process. Once the scheme is within budget the refined elemental estimate provides a cost control document. The cost of each element can then be monitored as the design develops and thereby prevented from unintentionally becoming unduly expensive or over-designed in comparison to other elements.

The employer is now in possession of an outline scheme which meets the basic parameters of the brief together with a cost plan based on known levels of quality. A decision as to whether or not to proceed can now be made. Beyond this point a commitment to progress the scheme will engender significant financial outlay.

Procurement

The method of choosing the contractor to construct the building must now be determined as this may dictate the manner in which the detailed design is progressed. For example, if a traditional procurement route is being followed the design team will develop the design, whereas with design and build procurement the design team may only prepare a design brief, the design itself being completed by the contractor. Each procurement option will have a different time, cost and quality scenario (the procurement triangle). These three elements may be held in a particular balance as the circumstances dictate. However, one element must always give way to the others and altering any one element will have an effect on the others – see Figure 3.1.

One interpretation of the time, cost and quality priorities which may be derived from three different procurement routes is given in Example 3.1. It must be stressed that the particular circumstances of each project may result in different conclusions as to the procurement route that best satisfies the priorities of time, cost or quality. For example, a high level of provisional sums within a contract could undermine the cost certainty otherwise afforded by a particular procurement route but would enable work to be started sooner if time is a higher priority. Additionally, it should be noted that for ease of presentation time allocation for building control approval has been omitted from Example 3.1. This element in itself can influence the choice of the procurement route.

It is advisable to apprise the employer of the procurement options, recommending the most appropriate route, detailing how the priorities are protected and outlining all implications. Only when the method of procurement is determined can the pre-contract programme be finalised. This also represents the point in time when the specific services required of each member of the design team can be defined, fee proposals confirmed and agreements finalised.

Detail design

The development of the sketch scheme into a workable solution will produce the detail design to the employer's brief and allow it to be submitted for detailed development control approval. The onus is on the design team to maintain the priorities established within the employer's brief and to extract and agree any further information required from the employer as the scheme progresses. Whatever the scale of the project, members of the design team have a responsibility to use their skills to:

- produce a good solution to meet their employer's brief
- consider end and/or future users if different from the employer
- design to meet the anticipated lifespan of the building to avoid early deterioration or costly maintenance
- include, as far as reasonably practicable, health and safety considerations in the design
- co-ordinate structure, finishes and services into one complete design.

During this development stage it is wise for the designer to hold preliminary meetings with the local authority to assess the likelihood of development control approval, to define any agreements the employer may have to enter into with the authority and to determine the timescale involved. If reaction is unfavourable or objections from others seem likely, the designer owes a duty to the employer to advise accordingly, indicating the possible extent of negotiation that may be required to achieve approval or the implications of appealing with regard to time, cost and risk of refusal.

In submitting a detailed development control application the design team is likely to have to produce the following information:

- floor layout plans
- typical sections showing proposed heights
- all elevations
- elevations or pictorial views, where relevant showing the proposals in context with adjoining buildings
- site plans showing relationship of the proposal to other buildings, orientation, access, parking standards and the like
- details of materials and colours.

This information should be presented legibly and accurately in view of the different groups, committees, departments and members of the public that may have the right or duty to give their opinion.

If the application is refused or the application is not determined within an agreed timescale or the employer objects to certain conditions imposed in an approval, the applicant has the right to appeal within a time limit. Great care should be taken in advising an employer on these matters, especially in view of the penalties imposed for unreasonable or unsuccessful appeals. The employer may be well advised to seek expert legal advice prior to pursuing an appeal, particularly if the proposal is complex or risky.

Once a detailed approval is granted an employer can instruct the design team to proceed to the production stage; if such instruction is given before the approval is granted the employer must accept the risk of abortive work.

Programming

When all submissions for statutory and other approvals have been satisfactorily completed and when the final design proposals have been given, the design team can re-examine the outline programme made at the start of the project and amend it in the light of the stage reached, the procurement option chosen and any financing conditions imposed.

The purpose of the detailed pre-contract element of the programme is to set out a sensible and logical sequence of the various pre-contract operations appropriate to the members of the design team and the required level of their input, together with external factors peculiar to the project. The initial bar chart or network can be expanded to show a programme to include:

- design team progress meetings
- preparing co-ordinated production information
- obtaining statutory approvals
- finalising legal agreements for access arrangements, party wall awards and the like
- negotiating with the service providers
- receiving and integrating specialist design elements
- receiving quotations from sub-contractors and suppliers
- finalising information for the pre-tender health and safety plan
- preparing contract conditions and preliminaries, including sequencing and contingencies
- preparing tender pricing documents
- preparing pre-tender estimate and updated cash flow prediction
- completion of pre-tender enquiries
- provision of tender documents
- receiving, appraising and reporting on tenders
- assembling contract documents
- briefing site inspectorate
- naming sub-contractors and suppliers.

In conjunction with these key points, the team should build in a contingency to cover such activities as co-ordination within the team, the inevitable drawing changes during the process and checking completed documentation. For the programme to run smoothly, all team members should be fully aware of the programme in respect of their own role, the interdependence of members of the design team and the effect that delay by one member will have on the others. It is the duty of the team to keep the employer informed of progress against the programme and to give advice on any factors which may alter the programme and their effect in cost and/or time.

Design team meetings

Chapter 2 refers to the traditional role of the architect as project leader. However, increasingly with larger projects, employers appoint a specific project manager to co-ordinate the pre-contract and site works on their behalf. A project leader should seek to manage the effective production of all the required information and ensure that the design team is working cohesively towards the common goal. The project leader should be aware, therefore, of the terms of appointment and limits of responsibility of each team member and ensure that they work together to produce an efficient and coherent design.

It is easy for time to be wasted if the members of the design team are not in regular contact. The project leader should hold regular design team meetings to review:

- new and revised information
- the detailed design
- the integration of the structural and services elements
- specialist design input
- health and safety principles
- progress against the programme.

Clear decisions must be made at these meetings. Time spent constantly changing drawings is unproductive and likely to lead to abortive costs. A record of decisions taken and any action required by a member of the design team should be made so that such actions can be checked off at subsequent meetings.

It is important that members of the design team should not lose sight of the fact that their role is to provide a service to the employer. Many professional practices operate strict quality assurance procedures to ensure the maintenance of good practice within their organisation and the provision of a quality service to any employer.

Drawings

The most traditional and widespread method of conveying fundamental information for building is the drawing. It is common for the architect to use computer and overlay draughting techniques and to commence by producing basic plans and sections (general arrangements) on which can be superimposed:

- setting-out dimensions
- structural engineer's designs
- services engineers' designs
- specialists' designs
- information for development control submissions.

Once setting-out, structure and services requirements have been determined, the design team may then proceed to the production of large-scale details.

The clarity of drawings and other documents is fundamental to the smooth running of any project. Hence the goal of the design team must be to make the project information as simple and as easy to understand as possible. It is suggested that the most effective way of achieving this is to adopt the conventions of Co-ordinated Project Information (CPI) as set out in a guide first published by the Co-ordinating Committee for Project Information in 1987.

Specifications

At the beginning of the pre-contract programme the design team, in conjunction with the employer, will need to make an early selection of the options for the appropriate performance requirements and/or the type and quality of all materials, goods and workmanship necessary to complete the project. When selecting specific materials and goods the design team will have to ensure that they:

- are appropriate to function, exposure and predicted use
- are available without undue difficulty
- provide appropriate levels of thermal resistance, sound attenuation and fire resistance
- comply with relevant health and safety conventions
- are environmentally friendly and energy efficient in manufacture and use
- create the required ambience in terms of space, colour and texture
- are easy to clean and maintain
- are appropriate to the employer's budget.

When translated into written form, the performance requirements and/or the qualitative details of the selected materials, goods and workmanship comprise the specification. Chapter 18 explains in more detail the different types of specification and their function.

Bills of quantities

The decisions on specification made by the design team and the employer form the basis of the written contract documentation. The method for presenting this

'... *early selection* ...'

information will depend upon the procurement route adopted. One of the more traditional procurement routes will require the production of bills of quantities. This process is one by which the quantity surveyor analyses the drawings and specification and, by following a standard set of rules, translates them into a schedule of quantified, descriptive items building up the constituent parts of the proposed project.

The primary purpose of the bills of quantities is to provide a uniform basis for competitive lump sum tenders and a schedule of rates for pricing variations. Chapter 19 explains in more detail the different types of bills of quantities and their functions.

Specialist sub-contractors and suppliers

The use of certain materials, goods or installations requires specialist knowledge and skills that are likely to be beyond those possessed by the main contractor. When this occurs the design team will normally approach specialist suppliers or sub-contractors for the provision of design and quotations for the work.

Quality assurance

Throughout all procedures the quality of the product or service should be at an acceptable level. Quality assurance is a management process designed to provide a high probability that the defined objective of the product or service will be achieved. To achieve quality consistently an organisation needs to have a suitable management system in place. That system should be capable of applying appropriate management checks throughout all work stages, be it a design service, a product manufacturing process or a building erection process, and of correcting deficiencies if they occur.

ISO 9000 is a series of standards for quality management systems overseeing the production of a product or service. ISO 9000 was originally published by the British Standards Institution under the guise of BS 5750, but is now maintained by the International Organisation for Standardisation (ISO).

ISO 9000 is a generic standard that can be applied to any company wishing to create a quality management system, whether it is large or small, for-profit or governmental, whatever the product or service. ISO 9000 standards focus on a number of principles, including a focus on the customer, sound leadership practices, involvement of people at all levels, a process approach, a systematic approach to management, continual improvement, a factual approach to decision making, and mutually beneficial supplier relationships.

A recognised way for a firm to set up its own quality assurance system is to prepare a quality manual covering:

- the overall policy of the firm as regards quality of service
- policies on such matters as information services, staff training, resource control and documentation
- the preferred methods of running projects, such as those based on the *Architect's Plan of Work* (RIBA Publications) or some other procedures manual.

The principals of the firm should formulate, endorse and evaluate the manual and communicate its contents to all personnel, ensuring full understanding. A firm may then decide whether or not it wishes to seek third-party assessment by a certification body leading to registration under ISO 9001. Such a body will first examine whether the manual prepared by a firm meets all the requirements of the International Standard and then whether the firm is operating in accordance with the quality manual. After registration, further third-party assessments are carried out at regular intervals to ensure that the firm continues to operate in accordance with its quality manual.

Quality assurance on its own will not improve a firm's professional standards: it only provides an auditable record of performance against the firm's stated objectives, which may or may not embody high professional standards. The adoption of quality assurance will promote consistency in performance only at the level set down within the quality manual.

Obtaining tenders

This chapter has discussed the various elements involved in the pre-contract programme and factors important to the smooth running of that programme. It has also examined the different types of information that are used to convey to the contractor the work to be done and to obtain tender prices. In Chapters 16 to 20 these are discussed in greater detail and Chapter 21 addresses the procedure for obtaining selective tenders.

Chapter 16
Pre-Contract Cost Control

Cost control

To enable employers satisfactorily to fund their desired building projects they will need to know how much each is going to cost them and when they will have to pay. The service by the design team of providing this information is referred to in this book as cost control.

Ideally, cost control should be provided from inception to completion of any project so that the current estimated final cost is always known. Pre-contract cost control in terms of approximate estimates, cost plans, life cycle costing and cash flow predictions is considered in this chapter. Post-contract cost control is dealt with in Chapters 26 and 27 and capital allowances are dealt with in Chapter 32.

Reporting

If the design team is to obtain maximum benefit from pre-contract cost control it is essential that a sound cost reporting regime be adopted. The quantity surveyor should provide cost reports in accordance with the plan of work (see Example 16.1). Also, because the level of information available on which to base the estimated costs will vary from time to time, it is essential that the quantity surveyor sets down in clear and concise terms as much of the following information as is available, always stating where assumptions have been made:

- date of report
- names of the design team members
- address of the site
- extent of the site and any restrictions (e.g. as to use or access)
- planning situation
- basis of costs
- anticipated rate of inflation relating to costs used

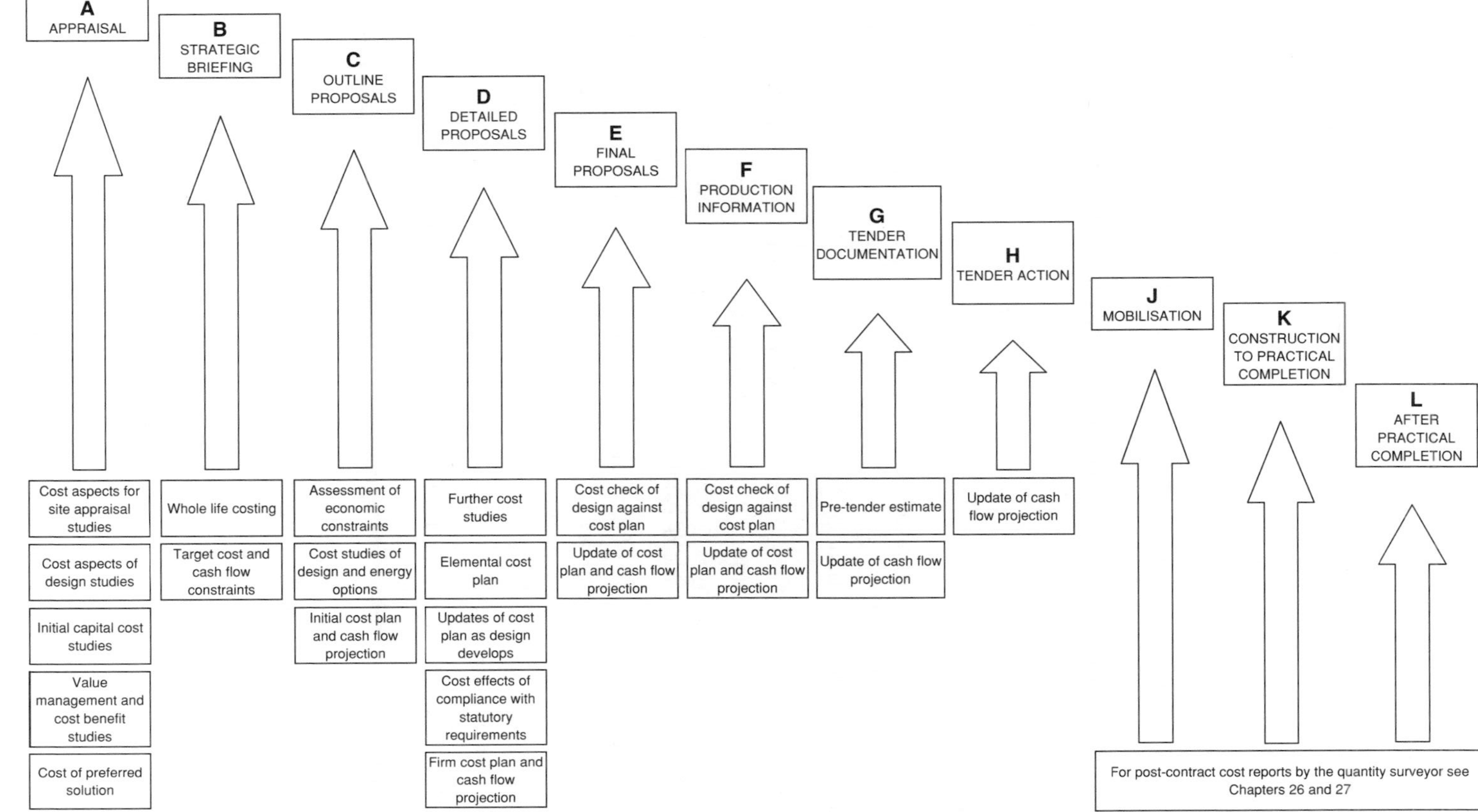

Example 16.1 Plan of work – pre-contract cost reports by the quantity surveyor.

- site investigation situation (e.g. site survey, ground investigation)
- requirement for any enabling works (e.g. demolitions, ground improvement)
- method of procurement
- date of commencement of the works on site
- contract period
- any phasing of possession
- any sectional completion
- schedule of drawings used
- outline of specification used
- availability of supply services (e.g. sewers, water, electricity, gas, telephone)
- exclusions (e.g. fitting out, professional and other fees, VAT)
- items with long delivery periods necessitating advance ordering
- a programme.

Establishing a budget

The detailed procedures to be adopted in pre-contract cost control will be governed, amongst other things, by the way in which the budget is established. There are three primary ways in which it is usual for this to be done:

- **Prescribed cost limit** The limit of capital expenditure is set down in the employer's brief, the amount normally being based on the maximum amount that can be afforded. 'Once in a lifetime' employers, not building for profit, will normally adopt this method of establishing a budget.

- **Unit cost limit** The limit of capital expenditure is calculated by the application of standard costs on a unit basis (e.g. £x per person to be accommodated, £y/m^2 of gross floor area to be provided). Public bodies and other employers who frequently procure buildings of a repetitive functional nature normally adopt this method of establishing a budget.

- **Determined cost limit** The limit of capital expenditure is based upon the economic viability of the proposed development. The employer will calculate how much of the non-recurring development costs are to be allocated to building costs if the scheme is to be economically viable. There is some latitude in this calculation, as the standard of building will, in part, determine the income to be derived from the scheme. All those building for profit will normally adopt this method of establishing a budget.

Approximate estimates

The object of the exercise is to estimate the cost to the employer of the desired scheme. It is usual to base the pricing of such estimates on historical cost data gleaned from the analyses of the costs of similar buildings adjusted to account for any fluctuations in price levels and for project idiosyncrasies.

The accuracy of the estimate will be influenced by the state of the design information, both drawings and specification notes. A 'preliminary estimate' is often required when there is little or no information available, so it is usual to base such an estimate on functional unit costs or net area/volume costs. As the design information is developed it is possible to produce more refined estimates based upon 'element unit quantities' or measured 'approximate quantities'. Finally, when the design is complete, an even more accurate estimate can be achieved by basing it on the measured 'accurate quantities' in the bills of quantities.

Preliminary estimate

As the whole economics of a project are frequently based upon the preliminary estimate it is not surprising that, of all the pre-contract cost information provided to the employer, the figures given at this time are the ones that will remain most clearly in mind even though, by their very nature, they could be the least accurate. It may well be that the preliminary estimate is required before any drawings, even simple sketch plans, have been prepared. Whether or not this is the case, it should not be produced until information on the following is to hand:

- the site and its nature
- the type of building and its use
- the overall height, number of floors and total floor area of each building
- an indication of the quality of materials and workmanship to be specified
- an outline of the engineering services to be provided.

The preliminary estimate will be based either on the units of accommodation or on the net floor area/volume to be provided priced at appropriate £/unit, £/m^2 or £/m^3 rates.

After completion of the preliminary estimate it will probably be necessary to carry out comparative cost studies of such matters as alternative plan shapes, numbers of floors, types of structure and other matters fundamental to the design.

Element unit quantities estimate

As more design information becomes available, a more refined estimate may be prepared to check the accuracy of the preliminary estimate. The proposed buildings will be analysed in terms of element unit quantities (see the checklist of elements given later in this chapter) to which elemental unit rates are applied. The estimated cost of the external works and services will be based upon individually priced approximate quantities (see below). An element unit quantities estimate should not be produced until the following are available:

- Sketch plans:
  - 1:200 plan of each floor
  - 1:200 elevations
  - 1:200 sections
  - 1:500 or 1:200 site plan showing extent of external works and location of external services.
- Brief notes relating to the extent and/or standard of:
  - substructure
  - frame
  - upper floors
  - roof
  - stairs
  - external walls
  - windows and external doors
  - internal walls and partitions
  - internal doors
  - wall, floor and ceiling finishes
  - fittings and furnishings
  - internal services (e.g. sanitary fittings, disposal, water, heating, electrical, gas, lift, communications)
  - external services (e.g. disposal, water, electric, gas, telephone)
  - external works (e.g. roads, pavings, walls, fences, gates, water features).
- Brief notes on any unusual conditions of work or contract.

Approximate quantities estimate

This is usually produced if the proposed construction is unusual and/or complex or the project involves the alteration of existing building(s). This type of estimate is prepared on the basis of individually priced approximate quantities that are inclusive of all labours. Only the main items of work need be measured. Wherever possible items having similar measurements, for example hardcore bed, blinding, damp proof membrane, insulation, reinforcement and concrete slab, can be grouped together. Relatively low-cost complicated elements of work can be itemised and individually priced. In many instances a building can be analysed into as few as 100 measured items representing something of the order of 90% of the total cost. Provided that care is taken in booking the measurements and in choosing the rates the approximate quantities estimate should be reasonably accurate. The information required is the same as that for the element unit quantities estimate plus:

- 1:100 plans of each floor
- 1:100 elevations of each face
- 1:100 sections
- typical details of important features.

Accurate quantities estimate

Such estimates are produced only when there are bills of quantities. These are priced before tenders are received (also referred to as a 'pre-tender estimate', or PTE) to check the accuracy of previous estimates, to give advance warning of where adjustments in design may be necessary to reduce costs and to provide a base against which to compare the lowest tender.

Cost plan

When members of the design team are satisfied that they have determined the manner in which the employer's requirements can be best met, sketch plans and approximate estimates can be produced for the whole project and a cost plan can be set up as a prelude to the preparation of working drawings, if this has not been done in conjunction with the earlier cost studies.

The cost plan is the document that brings design and cost together at pre-tender stage. When properly prepared and used it can assist the design team in both controlling the total building cost and spreading that cost between the various elements of the building in the most efficient manner, thus enabling the best use to be made of the employer's money.

...the best use of the employer's money

In order to prepare a cost plan it is necessary to divide the building into a series of elements (e.g. frame, roof, upper floors). The choice of elements is determined by the nature of the building and the services and features in it. However, in order to maintain consistency in the compilation of cost data it is essential to always use the same basic elements. Perhaps the best way of achieving this is to adopt the listing of elements that has been used for many years by the Building Cost Information Service. This is reproduced below and provides a comprehensive checklist of elements and their sub-divisions.

(1) Substructure

All work below underside of screed or where no screed exists to underside of lowest floor finish including damp-proof membrane, together with relevant excavations and foundations.

(2) Superstructure

2.A Frame

Loadbearing framework of concrete, steel or timber.
Main floor and roof beams, ties and roof trusses of framed buildings.
Casing to stanchions and beams for structural or protective purposes.

2.B Upper floors

Upper floors, continuous access floors, balconies and structural screeds (access and private balconies each stated separately), suspended floors over or in basements stated separately.

2.C Roof

2.C.1 Roof structure
Construction, including eaves and verges, plates and ceiling joists, gable ends, internal walls and chimneys above plate level, parapet walls and balustrades.

2.C.2 Roof coverings
Roof screeds and finishings.
Battening, felt, slating, tiling and the like.
Flashings and trims.

Insulation.
Eaves and verge treatment.

2.C.3 Roof drainage
Gutters where not integral with roof structure, rainwater heads and roof outlets. (Rainwater downpipes to be included in 'Internal drainage' (5.C.1).)

2.C.4 Rooflights
Rooflights, opening gear, frame, kerbs and glazing.
Pavement lights.

2.D Stairs

2.D.1 Stair structure
Construction of ramps, stairs and landings other than at floor levels.
Ladders.
Escape staircases.

2.D.2 Stair finishes
Finishes to treads, risers, landings (other than at floor levels), ramp surfaces, strings and soffits.

2.D.3 Stair balustrades and handrails
Balustrades and handrails to stairs, landings and stairwells.

2.E External walls

External enclosing walls including those to basements but excluding items included with 'Roof structure' (2.C.1).
Chimneys forming part of external walls up to plate level.
Curtain walling, sheeting rails and cladding.
Vertical tanking.
Insulation.
Applied external finishes.

2.F Windows and external doors

2.F.1 Windows
Sashes, frames, linings and trim.
Ironmongery and glazing.

Shop fronts.
Lintels, sills, cavity damp-proof courses and work to reveals of openings.

2.F.2 External doors
Doors, fanlights and sidelights.
Frames, linings and trims.
Ironmongery and glazing.
Lintels, thresholds, cavity damp-proof courses and work to reveals of openings.

2.G Internal walls and partitions

Internal walls, partitions and insulation.
Chimneys forming part of internal walls up to plate level.
Screens, borrowed lights and glazing.
Moveable space-dividing partitions.
Internal balustrades excluding items included with 'Stair balustrades and handrails' (2.D.3).

2.H Internal doors

Doors, fanlights and sidelights.
Sliding and folding doors.
Hatches.
Frames, linings and trims.
Ironmongery and glazing.
Lintels, thresholds and work to reveals of openings.

(3) Internal finishes

3.A Wall finishes

Preparatory work and finishes to surfaces of walls internally.
Picture, dado and similar rails.

3.B Floor finishes

Preparatory work, screed, skirtings and finishes to floor surfaces excluding items included with 'Stair finishes' (2.D.2) and structural screeds included with 'Upper floors' (2.B).

3.C Ceiling finishes

3.C.1 Finishes to ceilings
Preparatory work and finishes to surface of soffits excluding items included with 'Stair finishes' (2.D.2) but including sides and soffits of beams not forming part of a wall surface.
Cornices, coves.

3.C.2 Suspended ceilings
Construction and finishes of suspended ceilings.

(4) Fittings and furnishings

4.A Fittings and furnishings

4.A.1 Fittings, fixtures and furniture
Fixed and loose fittings and furniture including shelving, cupboards, wardrobes, benches, seating, counters and the like.
Blinds, blind boxes, curtain tracks and pelmets.
Blackboards, pin-up boards, notice boards, signs, lettering, mirrors and the like.
Ironmongery.

4.A.2 Soft furnishings
Curtains, loose carpets or similar soft furnishing materials.

4.A.3 Works of art
Works of art if not included in a finishes element or elsewhere.

4.A.4 Equipment
Non-mechanical and non-electrical equipment related to the function or need of the building (e.g. sports equipment).

(5) Services

5.A Sanitary appliances

Baths, basins, sinks, etc.
WCs, slop sinks, urinals and the like.
Toilet-roll holders, towel rails, etc.
Traps, waste fittings, overflows and taps as appropriate.

5.B Services equipment

Kitchen, laundry, hospital and dental equipment and other specialist mechanical and electrical equipment related to the function of the building.

5.C Disposal installations

5.C.1 Internal drainage
Waste pipes to 'Sanitary appliances' (5.A) and 'Services equipment' (5.B).
Soil, anti-siphonage and ventilation pipes.
Rainwater downpipes.
Floor channels and gratings and drains in ground within buildings up to external face of external walls.

5.C.2 Refuse disposal
Refuse ducts, waste disposal (grinding) units, chutes and bins.
Local incinerators and flues thereto.
Paper shredders and incinerators.

5.D Water installations

5.D.1 Mains supply
Incoming water main from external face of external wall at point of entry into building including valves, water meters, rising main to (but excluding) storage tanks and main taps.
Insulation.

5.D.2 Cold water service
Storage tanks, pumps, pressure boosters, distribution pipework to sanitary appliances and to services equipment.
Valves and tanks not included with 'Sanitary appliances' (5.A) and/or 'Services equipment' (5.B).
Insulation.

5.D.3 Hot water service
Hot water and/or mixed water services.
Storage cylinders, pumps, calorifiers, instantaneous water heaters, distribution pipework to sanitary appliances and services equipment.
Valves and taps not included with 'Sanitary appliances' (5.A) and/or 'Services equipment' (5.B).
Insulation.

5.D.4 Steam and condensate

Steam distribution and condensate return pipework to and from services equipment within the building including all valves, fittings, etc.

Insulation.

5.E Heat source

Boilers, mounting, firing equipment, pressurising equipment, instrumentation and control.

ID and FD fans, gantries, flues and chimneys, fuel conveyors and calorifiers.

Cold and treated water supplies and tanks, fuel oil and/or gas supplies, storage tanks, etc., pipework (water or steam mains), pumps, valves and other equipment.

Insulation.

5.F Space heating and air treatment

5.F.1 Water and/or steam

Heat emission units (radiators, pipe coils, etc.), valves and fittings, instrumentation and control and distribution pipework from 'Heat source' (5.E).

5.F.2 Ducted warm air

Ductwork, grilles, fans, filters, etc., instrumentation and control.

5.F.3 Electricity

Cable heating systems, off-peak heating systems, including storage radiators.

5.F.4 Local heating

Fireplaces (except flues), radiant heaters, small electrical or gas appliances, etc.

5.F.5 Other heating systems

Air treatment:

Air treated locally.

Air treated centrally.

Combination of treatments and whether inlet, extract or re-circulation.

High velocity system.

5.F.6 Heating with ventilation (air treated locally)

Distribution pipework, ducting, grilles, heat emission units including heating calorifiers except those which are part of 'Heat source' (5.E), instrumentation and control.

5.F.7 Heating with ventilation (air treated centrally)
All work as detailed under (5.F.6) for system where air treated centrally.

5.F.8 Heating with cooling (air treated locally)
All work as detailed under (5.F.6) including chilled water systems and/or cold treated water feeds.
(The whole of the costs of the cooling plant and distribution pipework to local cooling units shall be shown separately.)

5.F.9 Heating with cooling (air treated centrally)
All work detailed under (5.F.8) for system where air treated centrally.

5.G Ventilating system

Mechanical ventilating system not incorporating heating or cooling installations, including dust and fume extraction and fresh air injection, unit extract fans, rotating ventilators and instrumentation and controls.

5.H Electrical installations

5.H.1 Electric source and mains
All work from external face of building up to and including local distribution boards, including main switchgear, main and sub-main cables, control gear, power factor correction equipment, stand-by equipment, earthing, etc.

5.H.2 Electric power supplies
All wiring, cables, conduits, switches, etc., from local distribution boards, etc., to and including outlet points for the following:

— general-purpose socket outlets
— services equipment
— disposal installations
— water installations
— heat source
— space heating and air treatment
— gas installation
— lift and conveyor installations
— protective installations
— communication installations
— special installations.

5.H.3 Electric lighting
All wiring, cables, conduits, switches, etc. from local distribution boards and fittings to and including outlet points.

5.H.4 Electric lighting fittings
Lighting fittings including fixing.
(Where lighting fittings supplied direct by client, this should be stated.)

5.I Gas installations

Town and natural gas services from meter or from point of entry where there is no individual meter.
Distribution pipework to appliances and equipment.

5.J Lift and conveyor installations

5.J.1 Lifts and hoists
The complete installation including gantries, trolleys, blocks, hooks and ropes, pendant controls and electrical work from and including isolator.

5.J.2 Escalators
As detailed under 5.J.1.

5.J.3 Conveyors
As detailed under 5.J.1.

5.K Protective installations

5.K.1 Sprinkler installations
The complete sprinkler installation and CO_2 extinguishing system including tanks, control mechanism, etc.

5.K.2 Fire-fighting installations
Hose-reels, hand extinguishers, fire blankets, water and sand buckets, foam inlets, dry risers (and wet risers where only serving fire-fighting equipment).

5.K.3 Lightning protection
The complete lightning protection installation from finials and conductor tapes to and including earthing.

5.L Communication installations

The following installations shall be included:

5.L.1 Warning installation (fire and theft)
Burglar and security alarms.
Fire alarms.

5.L.2 Visual and audio installations
Door signals.
Timed signals.
Call signals.
Clocks.
Telephones.
Public address.
Radio.
Television.
Pneumatic message system.

5.M Special installations

All other mechanical and/or electrical installations (separately identified) which have not been included elsewhere, e.g. chemical gases, medical gases, vacuum cleaning, window cleaning equipment and cradles, compressed air, treated water, refrigerated stores.

5.N Builder's work in connection with services

Builder's work in connection with mechanical and electrical services (5.A to 5.M each item separately identified).

5.O Builder's profit and attendance on services

Builder's profit and attendance in connection with mechanical and electrical services (5.A to 5.M each separately identified).

(6) External works

6.A Site works

6.A.1 Site preparation
Clearance and demolitions.
Preparatory earth works to form new contours.

6.A.2 Surface treatment
The cost of the following items shall be stated separately if possible:

— roads and associated footways
— vehicle parks
— paths and paved areas
— playing fields
— playgrounds
— games courts
— retaining walls
— land drainage
— landscape work.

6.A.3 Site enclosure and division
Gates and entrance.
Fencing, walling and hedges.

6.A.4 Fittings and furniture
Notice boards, flag poles, seats, signs.

6.B Drainage

Surface water drainage.
Foul drainage.
Sewerage treatment.

6.C External services

6.C.1 Water mains
Main from existing supply up to external face of building.

6.C.2 Fire mains
Main from existing supply up to external face of building.
Fire hydrants.

6.C.3 Heating mains
Main from existing supply or heat source up to external face of building.

6.C.4 Gas mains
Main from existing supply up to external face of building.

6.C.5 Electric mains
Main from existing supply up to external face of building.

6.C.6 Site lighting
Distribution, fittings and equipment.

6.C.7 Other mains and services
Mains relating to other service installations (each shown separately).

6.C.8 Builder's work in connection with external services
Builder's work in connection with external mechanical and electrical services, e.g. pits, trenches, ducts, etc. (6.C.1 to 6.C.7 each separately identified).

6.C.9 Builder's profit and attendance on external mechanical and electrical services
(6.C.1 to 6.C.7 each separately identified.)

6.D Minor building work

6.D.1 Ancillary buildings
Separate minor buildings such as sub-stations, bicycle stores, horticultural buildings and the like, inclusive of local engineering services.

6.D.2 Alterations to existing buildings
Alterations and minor additions, shoring, repair and maintenance to existing buildings.

Development of the cost plan

Each element will have a particular design and/or specification characteristic and, although elements can influence each other, each will have a high percentage of its cost determined by design factors and the selection of material related directly to that element. Each element can therefore be considered in isolation until design, specification and costs are acceptable before its influences on other elements need to be examined.

The accuracy of the first cost plan will depend upon when and how it is prepared. If it is compiled at a very early stage by pricing element unit quantities (e.g. y/m^2 of external wall, x/number of internal doors) at rates taken from a similar earlier project, it is likely to be less accurate than if it is based on an elemental analysis of a project-specific approximate quantities estimate prepared at a later stage in the design process.

Any number of alternative design and specification solutions can be considered for each element. The cost of each should be estimated and superimposed on the cost plan to enable the design team to see how the various solutions affect cost. As the design is developed and decisions are made the cost plan must be updated. Where the value of any element is altered significantly by design development decisions, corresponding financially inverse adjustments will have to be made to other elements if the project estimated cost is to be maintained within the established budget. For this to happen it is essential that there is close co-operation by and good communication between all members of the design team throughout the design process.

By the time tenders are due to be received, the total of the cost plan should be a prediction of the lowest tender amount and should be at or below the established budget. If the lowest tender received exceeds the budget, the cost plan can be used to assist the design team in determining where savings can be made most effectively.

Life cycle costing

Although the appearance of a building is generally of great importance, functional factors can be of higher priority. A building may well be judged by the employer not only on aesthetics but also on the basis of the practicality of design, running costs, maintenance costs, its ability to be remodelled in order to allow changes in use, and the ease with which it can be extended. The design team must therefore ensure that the building will satisfy the user's requirements economically, not only initially but also for the whole life of the building. This will require a study of the effect of design and specification choices on the cost to the employer of the building in use.

Life cycle costing (or whole life costing) is a technique used to assist in the identification of the most economically efficient way of achieving the best value in terms of capital and revenue expenditure (cost in use) on a building, an element of a building or even a component within a building. The basic method is to use normal discounted cash flow techniques to evaluate capital expenditure and running costs. In this way the case for higher initial expenditure on a building, element or component can be evaluated against lower ongoing costs in use.

It is important when undertaking such studies to anticipate an appropriate, effective life for the building. There is little point in choosing expensive materials solely for their durability if they are likely to outlast the useful life of the building.

PROJECT NAME
for
EMPLOYER'S NAME

Prepared 6 September 2001

DATE	EVENT	CERTIFIED AMOUNT £	VAT @ 17.5% £	TOTAL PAYMENT £
07/01/02	Date of Possession			
07/02/02	Interim certificate 1 issued			
21/02/02	**Final date for payment of interim certificate 1**	**133,000**	**23,275**	**156,275**
07/03/02	Interim certificate 2 issued			
21/03/02	**Final date for payment of interim certificate 2**	**209,000**	**36,575**	**245,575**
08/04/02	Interim certificate 3 issued			
22/04/02	**Final date for payment of interim certificate 3**	**247,000**	**43,225**	**290,225**
07/05/02	Interim certificate 4 issued			
21/05/02	**Final date for payment of interim certificate 4**	**209,000**	**36,575**	**245,575**
24/05/02	Date for Completion			
07/06/02	Interim certificate 5 issued			
21/06/02	**Final date for payment of interim certificate 5**	**99,000**	**17,325**	**116,325**
24/11/02	End of Defects Liability Period			
24/01/03	Final Certificate issued			
28/02/03	**Final date for payment of Final Certificate**	**23,000**	**4025**	**27,025**
	TOTALS	**£920,000**	**£161,000**	**£1,081,000**

Example 16.2 Cash flow forecast for building work.

Cash flow

The capital cost of a building project will normally be funded from the employer's reserves or by loans from financial institutions. Employers will need to know the dates and amounts of payments in advance if they are to be able to safeguard against unnecessary loss or charge of interest. Employers could require forecasts of payments to various third parties, the most frequent of which are:

- landowner – for site purchase
- contractor – for building work
- statutory authorities – for mains services connections
- service providers – for electricity, gas and water
- local authorities – for planning and building control fees

- consultants – for professional services
- financial institutions – for insurances, interest and other funding charges.

See Example 16.2 for a very simple cash flow forecast for building work based on the cost plan at pre-tender stage. When the contract is let it will become necessary to amend the forecast so that it reflects the contract sum, the date of possession and the date for completion.

Chapter 17
Drawings and Schedules

The role of drawings and documents

Project risks for the employer, the architect and the contactor are reduced by the architect and design team achieving a set level of quality in the design documentation and by all parties using quality management processes. Above all, those who produce project information, when preparing the work, have to think carefully about the purpose of the particular set of drawings or documents and the users of the data. For example drawings and a description of a project aimed at securing outline planning consent will be only of passing interest to a contractor at tender stage.

In history drawings achieved extraordinary importance when they allowed the design function of the master mason to become detached. Drawings became the designer's representative on site, allowing the architect to be based remotely and work on a number of projects. The architect's role became defined independently of the craftsman and allowed unrestricted development. No longer was the architect obliged to be the combined designer/craftsman, based for long periods of time on a particular site. However, no longer could he provide constant design development, interpretation and experimentation on site on a day-by-day basis as the work progressed. Drawings produced an imperative to have ideas and technical solutions well thought and developed before issue. That imperative continues.

The general principles of architectural drawing have been developed over centuries and today's architects have inherited a long tradition. With developing drawing techniques they have striven always to communicate intent with increasing accuracy and completeness. The triadic system of plans, sections and elevations in use today was developed before the 13th century. Scalar drawing emerged around the 15th century, with constructed perspective drawings coming into use in the 16th century. The 17th century saw the development of projections, stereometry and descriptive geometry, followed in the 18th century by the codification of the triadic system, sciography and rendered surfaces. Axonometrics made an appearance in the 19th century. What of the 20th and 21st centuries? The need for even greater accuracy of project representation combined with the powerful calculation and representational powers of computing, has seen the strident development of computer aided design technologies. The ability to create complex data models of complex building forms, and further to assess their performance in terms of design experience, quality of internal environment, environmental impact, stability under extremes of weather and seismic conditions has transformed design practice. Clients and users can test potential building designs before committing to project proposals. Statutory authorities can also test projects against their own parameters. The ability to create highly realistic models has significantly reduced the risk of project uncertainty. Performance is predictable.

However computer aided design has brought its own problems. The software used to create designs using CAD is proprietary, highly complex and impenetrable. The question of an architect's liability for work done using CAD remains under discussion, although a key consideration is that the architect remains liable for his work and for the accuracy of drawings whether produced by pencil or computer.

Drawings and documents are not products. One issue is that repeated and uncontrolled use of design documents by a client may cause the documents to be treated as 'products' generated by the designer and may give rise to product liability risks. There are also issues concerned with the ownership, use and transmission of design data (or 'instruments of professional service') which could include contractors using CAD data never intended for that purpose, the integration of CAD data into facilities management programmes, or the reuse of electronic project data in projects far beyond the original purpose or well out of time.

Services such as 'delayering' information, using CAD data as the basis for an 'as built' survey set of documents, or allowing the data to be used as the basis for a facilities management programme or operations, may be regarded as additional services beyond the scope of the standard fee. The attitude currently prevailing is that electronic information is a component of the instruments of service and is only for the employer's benefit on a specific project and for specific use. Because of the rapidity of development in software and the hardware technologies used, designers need to be indemnified against any future issues arising out of compatibility, usability or readability of design information.

Quality

Drawings represent the most important means by which designers convey their intention to the rest of the building team and to statutory authorities. As stated in Chapter 15, the clarity of drawings is fundamental to the smooth running of any project. This cannot be over-emphasised. Clear, concise, well-planned, co-ordinated drawings not only make the information they contain easy to understand, they also inspire confidence. Conversely poor drawings do little except reveal the designer's lack of knowledge and inability to conduct affairs in a well-ordered manner.

The quality of information produced by the architect will either help or hinder the defence in any possible future legal action against him or for which he has to produce information for the defence of others. Recommendations on the preparation of drawings, especially working drawings, are contained in BS EN ISO 4157:1999. Also important are the principles defined in BS EN ISO 9001:2000 on quality assurance concerning the orderly preparation, dissemination and recording of design data and indeed all documentation relating to any form of manufacturing and production.

The method of producing drawings will vary from office to office depending on the size and resources of the practice. Some may rely heavily upon the use of computer aided design, others on traditional draughting techniques, but most will rely on a combination of the two. The type of employer, the nature of the project, the degree of involvement with other consultants and their resources and, finally, the programme for the work all determine how the drawings at each stage are to be dealt with.

But if quality is to be assured, there are simple rules that need to be observed no matter how the drawings are produced.

Quality manuals

An increasing number of professional firms have written quality manuals used to guide and control the quality and process of preparing documentation and conducting the professional and technical affairs of the firm. Quality manuals need to be assessed in action over a period of time by an independent organisation before the latest international standard, BS EN ISO 9001:2000, can be awarded. This accreditation provides clients with a degree of assurance that the firm follows specific procedures in the way it deals with client assignments. An increasing number of clients in both the public and private sectors require that firms indicate they are quality assured when they make proposals for work.

Quality manuals help firms to:

- provide an imaginative, competent and consistent professional service to their clients
- achieve, sustain and improve the quality of design services provided in a way that will consistently meet the stated, implied and perceived needs of clients
- provide the documented assurance to clients that the intended quality of service has been and will be achieved.

Quality procedure codes

Quality procedures call for documents to be coded. An example is a four-part code comprising:

- **Document code** To define the document type. These could be:
  - — MA Manual
  - — PR Procedure
  - — FM Form

— PN Practice note
— RR Register (for system control).

- **Responsibility code** To define the area of activity within the quality management system and those who are responsible for the writing, approving and distribution of system documents. These could be:

  — AD Administration
  — DM Design management and project control
  — SM System management and maintenance
  — MS Management system information.

- **Document numbering code** To identify the serial number of the document within its own document type and responsibility code series. These can be simple two-digit numeric references such as 01.

- **Status code** To identify the revision status of specific documents. It also distinguishes between documents that are still in draft format and those that have been brought into use. Examples could be:

  — OA First draft
  — OB Second draft et seq.
  — 01 First issue
  — 02 Second issue et seq.

Quality review

The procedures described below need to be established to review, audit, initiate both corrective and preventive action, provide internal feedback and record client satisfaction (or dissatisfaction):

- **Management review** Reviews of the quality system in the firm should be conducted by the quality manager at least twice yearly to evaluate its effectiveness and provide for continual improvement.

- **Internal audit** All processes within the firm should be subject to systematic and independent internal audit by suitably trained staff in accordance with audit programmes.

- **Corrective and preventive action** The cause of all non-conformity with procedures and standards is identified through:

  — risk assessment
  — audit
  — design review and verification
  — feedback
  — post-project review on selected projects.

The results should be analysed and appropriate corrective action implemented to prevent recurrence.

- **Internal feedback** A simple mechanism should be established to enable all staff to propose changes and improvements to the quality system based on experience. All information and proposals should be considered and action taken recorded.
- **Client satisfaction (or dissatisfaction)** The level of client satisfaction should be recorded on both on-going and completed projects. This is effected through internal audit during the progress of projects and post-project review on completed projects.

Types, sizes and layout of drawings

The types of drawings vary as the project proceeds through its programme, as described later in this chapter. However, for efficient and economic use of time all drawings should:

- be on standard-sized sheets laid out in such a manner that the source and purpose of the drawing can be readily identified
- contain all necessary routine information in a form that can be readily checked
- be kept in comprehensive sets and stored easily (record sets may have to be kept in archives for up to 25 years).

An assortment of drawings of different shapes and sizes with title panels and essential information in differing positions is a cause of confusion and irritation. Standardisation should apply no matter how the drawings are prepared.

Size

In BS EN ISO 5457:1999 recommendations are made for drawing sheet sizes A0, A1, A2, A3 and A4. In selecting the standard drawing sizes, account should be taken of the many and varied methods of reproduction: dye-line printing, photocopying, laser printing and photographic. Excluding routine dye-line printing, all methods available allow the possibility of changing the scale of the drawing in the course of reproduction – an invaluable asset in planning and producing comprehensive and co-ordinated sets of drawings. With this advantage in mind it is probable that the range of standard sheets might reasonably be limited to A0, A1 and A3.

A0 is a rather unwieldy size, especially for handling on site, but may be necessary for general arrangement drawings of large projects, townscape plans or extensive landscape drawings and the like. A1 size drawings are the most common and the most acceptable for ease of handling and can be reduced to any size down to A4. This is extremely useful in the preparation of design and presentation drawings that may be required in bound folders of A3 or A4 size. A3 sheets are most commonly used for quick sketches and details that may be reduced down to A4 size for use in transmission by fax.

With the almost universal use in offices of computers, the use of drawing sheets for the preparation of schedules is the exception rather than the rule. Standard letters, forms of instructions, certificates and most schedules are stored electronically for completion or amendment as the occasion arises. Drawing sheets might be used for window, door and ironmongery schedules that contain a considerable amount of information – in this case A3 sheets would be the most convenient. The vast majority of drawings for a number of projects are produced at A3 size.

Layout and revision

The design of drawings must take account of both immediate and long-term storage. Whether it is for reduction and storage in files, in plan chests or in large clips hung on racks beside working stations, the layout of the sheet must provide sufficient margin to ensure that no information is obscured by the storage system.

Two types of title and information panel are recommended in BS EN ISO 4157:1999. The examples include space for classification references, but this will not be required unless a classified form of specification is being used. The latest and most widely adopted classification is the Unified Classification for the Construction Industry (Uniclass). This is a comprehensive coded system providing an interface between the manufacturer's product database, the use of materials and methods of assembly shown on the drawings, the specification and the bills of quantities.

Seldom are drawings produced and left unchanged. The development of the design and collaboration between members of the design team both lead to a continuing process of amendment and revision. Every revision to a drawing must be adequately described and the date when it was made noted on the sheet and indicated by a unique revision code. If this is not done, it will be difficult, if not impossible, for other members of the building team to discover changes made to the drawings. Despite the widespread use of computers with cheap storage, the use of broadband for fast transmission of large amounts of data and the use of extranets, it is still necessary to store project documentation in paper form for both practical and legal reasons. This will include the long term storage of 'as built' information.

Scale

Most printing processes distort drawings – some are reduced in size arbitrarily for the convenience of storage or transmission by fax and some may be microfilmed for long-term storage of 'as built' information. It is therefore good practice to incorporate

...long term storage of 'as built' information

clearly the appropriate drawn scale(s) on every original drawing. In any event it must be remembered that it is always dangerous to take dimensions from any drawing by scaling.

Difficulties can arise if drawings are prepared to unusual scales. Engineering consultants frequently use the scales of 1:25 and 1:250. This can cause considerable confusion and should be discouraged. Another confusing scale is 'half size' as it is too easily mistaken for full size. It is recommended that the following scales be used:

Location plans

- 1:2500
- 1:1250
- 1:500

Site and development plans

- 1:500
- 1:200

General arrangement drawings

- 1:100
- 1:50

Component drawings

- 1:50
- 1:20

Details

- 1:5
- 1:1 (full size)

Assembly drawings

- 1:20
- 1:10

Nature and sequence of drawing production

The type of contract and the method of choosing the contractor can affect the nature of the drawings and the sequence of their production. The building consists of many elements – structural frame, walls, partitions, roof, mechanical services and so on. Each of these elements forms a part of the whole and only a complete set of drawings can inform the contractor on the complete building. However, that does not necessarily mean that all drawings need to be completed before the contractor starts work on site – hence the need at the outset to understand the type of contract to be used.

The production of the drawings should be carefully planned. A preliminary list should be prepared of the drawings that will be required, not only from the architect but also from the other consultants.

Drawings for SBC contracts

When tenders are to be obtained and a contract placed on the basis of the SBC (or indeed JCT 98 which is still in fairly common use) with or without bills of quantities, all the drawings, schedules and specifications need to be completed prior to obtaining tenders, except when the contractor is provided with an information release schedule which states what information the architect will release and the time of its release. Reference to the Outline Plan of Work (see Figure 2.1) will show how important it is to plan the production of the drawings, particularly through work stages E and F. It should also be borne in mind that where bills of quantities are used the SMM7R general rules set out details of the drawings required for the purposes of tendering.

The drawing production sequence is as follows:

- **General arrangement drawings** Plans and elevations (1:100 or 1:50) which, after receiving the employer's approval, are sent to all consultants who will then prepare their draft schemes.
- **Services drawings by the architect (plumbing, drainage)** These are worked out in detail concurrently with the preparation of the general arrangement drawings.

- **Construction details by the architect and specialist sub-contractors** The selection of materials and finishes made during the preparation of the general arrangement drawings leads to the preparation of the 'architect's details' and to obtaining tenders for specialist sub-contract design and construction. The outcome of this stage of the work can affect other aspects, particularly critical dimensions in the primary elements of the structure, and consequently the work of other consultants.
- **Assembly details** When the drawings of various consultants are accepted, the assembly details (1:20 and larger if necessary), which have been drafted in outline, should be completed.
- **Final co-ordination of drawings** When all the detailed information has been assembled and co-ordinated the final overall picture (originally the outline general arrangement drawings) can be completed with accuracy.
- **Layout and site plans** These are finally completed, incorporating information on all external services and the setting out of the buildings.

Many details and drawings will be changed during the course of the works and, in the process of adjusting the costs, it will be necessary to refer to the original information upon which the tender and contract sum were based. It is therefore of the utmost importance to keep a record set of the drawings used in the production of the other contract documents.

Drawings for design and build or management contracts

There are occasions when it is appropriate for the contractor to be responsible for the production of the building design or to manage a contract in such a manner as will accelerate the rate of construction (even though it might increase the cost). In such circumstances the preparation of the drawings will be geared not to the process of obtaining tenders for the whole of the works to be carried out by a single contractor but to obtaining tenders from sub-contractors for the various parts of the building, known as 'packages'.

Once the building is designed, the sub-contract packages involved will be identified. The general arrangement drawings will have the same significance as in the traditional procurement route, but, once they are agreed, subsequent drawings will be of an elemental nature, that is, concerned with the various and separate elements of the building. The tenders for these packages will be obtained on the basis of drawings and specifications, and sometimes bills of quantities, for each package in a sequence related to the programme of work to be carried out on site. The order in which the drawings are produced will be related to the order of procurement and construction. Pile or foundation drawings, together with the main structural frame and drainage drawings, will be completely detailed and the work started on site, possibly before detailed or assembly drawings have started for secondary elements of the superstructure. Needless to say, the co-ordination of details in these circumstances becomes very difficult and can result in a rather different approach to design, calling for greater flexibility in the use of the structure or the building envelope to more readily allow substantial changes to the design late in its development.

The following is an example of the order in which drawings, specifications and bills of quantities (when used) for packages might be required. Those items marked with an asterisk (*) typically have a long lead-in or delivery period:

- general arrangement drawings
- levelling and general site works
- structural frame (steelwork)*
- drainage
- foundations (including piling)
- lifts and escalators*
- proprietary walling*
- roofing systems
- windows and doors*
- brickwork details
- mechanical and electrical engineering services*
- plant and machinery details
- partitions
- staircases and other secondary elements
- ceilings
- fixtures and fittings
- external works.

There can be many more packages than are given here and some that are given separately may be combined.

Design intent information

New forms of contract have their own requirements for the production of project documentation. The recently introduced JCT MP contract imposes demanding design submission procedures to amplify the design undertaken by the contractor.

In certain circumstances the design team may be asked to produce design intent drawings. One example is that of a contractor working on a Private Finance Initiative (PFI) project as part of a consortium requiring information to pass to the in-house design team.

The paradox of design intent drawings is that in many instances they are virtually complete as they have to communicate specific designs, details and specifications to another professional team and will be used for value management and detailed costing purposes. They will also be reviewed by the end user's team to assess value for money in the PFI context.

It is important that the design team led by the architect determines, at the outset of the production information stage, the exact level of detail required to be given in the design intent set.

Computer aided design

The use of computers in architecture is now almost universal. The title given to this technique describes well the role of computers in the design process. Computers aid but they are not creative, rather they power-assist the creative process by enabling designers to assess their work accurately and rapidly. Computers are essential tools with many attractions to all the members of a design team. The main advantage lies in their ability to perform at great speed the technical services of producing drawings, of calculating and co-ordinating and ultimately of communicating. If there is a disadvantage, it may be seen in the type of drawings most commonly produced – of single line weight without enhancement or emphasis and often, for all their precision, lacking the special quality that allows easy reading and interpretation.

The most simple computer system will produce two-dimensional drawings containing no more information than is input for each drawing. The most advanced will store a complete model of the building and, on command, produce drawings to any scale or projection of any component part, section or the whole of the building. Moreover, amendments to the original model will automatically appear on all drawings as they are produced. The storage capacity of even modestly powerful computers can carry a huge library of standard components, assemblies and details that will enable the architect or technician quickly to produce a vast range of working drawings of great accuracy.

Where members of the design team are using compatible hardware, discs can be exchanged giving each designer immediate access to all the information produced by the others. Layer upon layer, information can be added to the general arrangement drawings, ensuring complete co-ordination of all the parts of the building – structure, envelope details, mechanical and electrical engineering services and finishing details.

At the outset of the design process, members of the design team should discuss and agree the management of their CAD resources. All members of the team should produce one or more back-up copies of all drawings at all stages of production. Records of drawings undergoing major amendment should be stored. Common systems for coding drawings might well be adopted. On large projects, where many thousands of drawings may be produced, it is possible for the designer to have his system linked to specialist agencies which are made responsible for cataloguing and distributing to members of the building team the many drawings as they are produced or revised.

CAD enables every conceivable calculation related to the design process to be carried out, including structural, thermal performance, energy management, fire safety and lighting. And by following the procedures of CPI (Co-ordinated Project Information), the designers' programs can be interfaced with those of the quantity surveyor so that coded information on the drawings can be used in the production of the bills of quantities, specification documents and ongoing cost estimates.

The use of CAD does not change the ground rules for good practice, but the systematic approach that it requires does help to improve the clarity of the information produced and makes it more accessible. Whichever way the drawings are produced, the information they convey should be the same.

Drawing file formats and translation

A wide range of CAD systems, with varying software and training costs, is now in use, responding to the needs of architecture, interior architecture and the various engineering disciplines. With this diversity of software and capability come associated problems of the quality of the interface between disparate systems. The .dxf format is one method of storing information for use across a wide range of software platforms. One simple and widely occurring example is exchanging data between Microstation and Autocad formats. However, this method is not infallible and certain levels of detail may be lost in the translation process.

At the job set-up stage of any project, the preferred CAD platform should be agreed, and if commonality cannot be achieved, a translation protocol should be devised. Design files should be checked carefully after translation to avoid errors of omission or distortion.

Project extranets

One of the more significant problems in the construction industry is that of the quality of communication between members of the building team. The traditional system is inherently inefficient. The employer was not part of the communication system and relied on maintaining a very close working relationship with the design team members to keep up with their work and the method of transmission of information between members of the building team had not really changed in centuries.

Project extranets have been developed to overcome the wide range of problems related to the access of members of the building team to the most up-to-date information. There are at least 170 extranet companies without commonality of format, although this volatile marketplace will, through failures and amalgamations, settle down to serve the industry better.

Some systems are stronger during the design process while others perform best during construction. A project extranet is established along the lines of an intranet, using the Internet to form an electronic project community. A file server holds all project information produced by all participants. This information can be accessed by members of the building team (including the employer) on an on-demand basis from any location. Some extranets provide facilities for WebCams, so that site progress can be remotely checked by everyone concerned on a real-time basis and from anywhere in the world. Project information is classified according to stage, status and relevance. Access and the ability to manipulate information are carefully controlled. Users do not need to install a plethora of application software on their machines to view output from many sources. Although an important factor to consider when choosing a project extranet is to ensure that the chosen system will accommodate all the platforms used by the members of the building team. Most systems claim to run on browser technology but some operate only on Windows, thus excluding users of Unix, Macintosh and Linux. The need for a common 'engine' for all project extranets is important.

Above all, a project extranet changes fundamentally the relationship between the members of the building team. From members being passive recipients of information received one from another, they become active participants, having to go to the extranet to extract information. This fundamental shift of responsibilities will improve the performance of the building team and ultimately the quality of service given to the employer.

Project extranets provide the members of the building team with the ability to:

- share and organise project documents, plans and files but also set user-definable access control at the project, folder and document levels
- access project information on a 24/7 basis in a secure, accessible, on-line location
- automate the creation and distribution of project updates and document revisions
- easily communicate with each other anywhere in the world
- monitor project status and progress
- access and view project audit trails and document version history
- red-line documents, issue, control and manage requests for information and add comments to or indicate changes on project drawings, contracts and other documents
- view on-line CAD output, photographs, spreadsheets, faxes and documents whilst maintaining data integrity
- automatically maintain relationships within compound documents
- access a complete document history, including copies of earlier document versions, revisions and access records
- easily find specific documents within very large projects, via full text search or by document attributes such as file, name, type and author
- access disaster recovery facilities for project data
- access project-cloning capabilities, enabling all projects to be organised in the same way and allowing the rapid implementation of new projects.

The use of extranets allows business process management techniques to be deployed in the design and construction process, with the following advantages:

- shortens project schedules
- reduces costs
- increases productivity
- extends best practice tools and techniques throughout a project
- reduces project risks
- establishes full project accountability.

Some JCT contracts allow electronic communication, and the parties to contracts can agree to use such paradigms by entering into an Electronic Data Interchange (EDI) Agreement. Two standard forms exist; the EDI Association Standard EDI Agreement and the European Model EDI Agreement.

Proprietary extranet software provides for structured permissions to read or alter design data. There are ever-present issues of the ownership, use and transfer of instruments of professional service, and the danger that contractors or clients regard design information as a 'product' with liability exposure. The reuse of data, the incorporation of design data into facility management programmes, and the general dangers inherent in the electronic transfer of design data are ever present. The delivery of a drawing must not carry with it an express warranty or guarantee of exactness of information. Hard copies of drawings and data, archived by the architect, will retain control over amendments.

Extranet case study – BAA plc

BAA plc is the world's largest airport operator and has been using an extranet for part of its annual capital development programme, through the UK partner of a major US extranet provider. With a capital investment programme of around £1 million a day, BAA is one of the UK's principal developers of infrastructure and one of the construction industry's largest clients.

BAA and its framework partners are using a common extranet on a variety of projects, including new-build, extensions to existing buildings, refurbishments and maintenance. BAA has introduced new collaborative extranet technology across its Heathrow, Gatwick and Stansted airports to:

- improve efficiency and the timely sharing of the correct information
- reduce errors and wastage
- facilitate the re-use of information
- reduce life cycle costs
- improve the capturing of knowledge.

Contents of drawings

When planning and programming the production of drawings, account must be taken of the fact that every part of the building and its site has to be designed, detailed and drawn. Obvious as this might seem, it is unfortunately only too common to find elegant sets of drawings repetitiously showing the simple but assiduously avoiding the complicated. Should a part of a building be difficult to detail and to draw, that is the very detail and drawing that the contractor will need most. In considering how all the necessary information is to be conveyed, there are good and simple rules to observe:

- Draw every part of the building and do not repeat information unnecessarily.
- Sections are invaluable – indicate and code their location clearly.
- Remember that plans are merely horizontal sections; one plan for each floor level may not be enough.
- Number all rooms and spaces.
- Give reference numbers to all doors, windows, built-in fittings and the like; show these on all plans, sections and elevations.
- Make clear cross-reference to other drawings or to schedules, so that detailed information can be traced.
- Be consistent in showing structural grids and levels on all plans, sections and elevations.
- Provide concise notes laid out clear of drawn information.
- Work out and indicate all essential dimensions but do not labour the obvious.

There follows a checklist of drawings and their contents.

Survey plan

- Existing site, boundaries and surrounds
- Positions of major features such as existing buildings, roads, streams, ponds, walls and gates
- Positions, girth, spread and type of trees
- Location and type of hedgerows
- Sufficient spot levels and contour lines, related to a specific datum, to enable a section of the site to be drawn in any direction
- Position, invert and cover or surface levels of existing drains
- Location, direction and depth of service mains
- Rights of way, bridle paths and public footpaths
- Access to site for vehicles
- Ordnance reference if available
- North point

- A key plan showing the relationship of the various sheets, should the survey cover more than one sheet.

Site plan, layout and drainage

- Relevant information from the survey plan
- Building profile and grid dimensionally related to a datum point or line
- New roads and paths with widths and levels marked
- Floor levels clearly indicated using the same datum as for the existing levels on the survey plan
- Steps and ramps where they occur
- Trees and hedgerows removed or retained
- Street and other external lighting
- Soil and surface water drains complete with pipe sizes and connections to sewers, making clear the distinction between soil and surface water drains and manholes (manhole sizes, levels and invert levels should be shown on a separate schedule)
- Gas, water and electric mains with depths indicated where possible and showing positions of connections to existing mains, supply company meters (external) and details of meter housings if required and points of termination within buildings
- Banking and cutting, and areas of disposing or spreading surplus soil
- Details of fencing, existing and new.

General arrangement

Once the design is approved and working drawings commenced, copy negatives should be taken of the general arrangement drawings of all floors showing only door swings in addition to walls and partitions and without any dimensions or other information on them. These drawings can then be developed without confusion or loss of time to show different details.

Services

- Incoming mains and meter positions for all services
- Electrical layout, including power points, light points, switches and all electrical equipment
- Heating layout showing boilers, calorifiers, radiators or similar heating equipment
- Plumbing and internal drainage layout
- Gas layout
- Sprinkler system layout, including pumps and storage tanks

- Fire detection and protection equipment
- Location of any special services.

Foundations

- The main building grid annotated and dimensioned
- Width and depth of all foundations for walls, piers and stanchions, with levels to underside
- Positions and levels of drains, gullies and manholes close to foundations
- Walls above foundations dotted with wall thicknesses dimensioned
- Positions of incoming service mains, service main ducts and trenches and their levels.

Floor plans

Complete plans drawn at a constant level through all openings for all floors and mezzanine floors with:

- Overall dimensions
- External dimensions, taking in all openings
- Internal dimensions to show the positions and thicknesses of internal walls, partitions and all major features
- Doors, their direction of swing and reference numbers
- Windows and their reference numbers
- Names or numbers of all rooms and circulation spaces (see BS 1192)
- Vertical and horizontal service ducts, holes in floor slabs, flues and builder's work in connection with services
- Staircases, their direction of rise and stair treads numbered
- Hatching to indicate materials of which walls and partitions are constructed
- Rainwater pipes
- Profile of joinery fittings and their code or reference numbers
- Sanitary fittings
- Floor levels clearly marked.

Roof plan

- Main construction features
- Levels clearly marked
- Types of covering
- Direction of falls

- Rainwater outlets, gutters and pipes
- Rooflights
- Tank rooms, trap doors, chimneys, ventilation pipes and other penetrations; builder's work in connection with services
- Parapets, copings and balustrades
- Duckboards, catwalks, escape stairs and ladders
- Lightning conductors, flag poles and aerials.

Elevations of all parts of the building

- New and old ground levels
- Profiles of foundations
- Damp-proof course levels
- Levels of ground and upper floor slabs
- External doors and windows with their reference numbers
- Air bricks and ventilators
- External materials, flashings
- Rainwater pipes and gutters
- Lightning conductors.

Descriptive sections

It is important to have sections that describe the juxtaposition of all elements of the building, adjacent buildings and the general site area. The sections illustrate:

- New and original ground levels showing cut and fill
- The main structural profile with grid lines
- Type and depths of foundations
- Floor and roof structures
- Windows, doors and rooflights with their reference numbers
- Lintels, arches and secondary structural members
- Damp-proof membranes and courses, flashings
- Eaves and valley gutters, parapets, rainwater heads and downpipes
- All vertical dimensions.

Ceiling plans at all floor levels

- Room names and numbers over which ceilings occur
- Type of ceiling, type of suspension, suspension pattern and finishes
- Height of ceiling above floor level

- Location of light fittings, their type and size
- Fire detection and safety installations, ceiling void compartmentation, smoke detectors, emergency lighting
- Mechanical ventilation registers
- Sprinkler layouts.

Construction details (scale 1:20 and 1:10)

- Detailed sections for external walls, foundations and roofs
- Plans, sections and elevations for staircases
- Lift wells and escalators
- Any room or part of the building the setting out of which is difficult, involves extensive fittings, fixtures, plumbing or special features or requires careful co-ordination, such as:
    - — kitchens
    - — bathrooms
    - — lavatories
- Special-purpose rooms in hospitals, laboratories, etc.
- Window and door details
- Part elevations and sections of the building's envelope containing special features such as:
    - — entrances
    - — special brick details
    - — balconies
    - — stonework
    - — plant rooms
- Builder's work in connection with services
- Vertical and horizontal pipe ducts and their access
- Fireplaces and flues
- External details such as special paving, steps, kerbs, handrails and external lighting.

Large-scale details (scale 1:10 and 1:15)

These comprise enlargements of component parts of assemblies which have been shown at 1:20 scale but which require a larger scale to show the full details:

- Sills, heads and jambs of windows and doors
- Special brickwork, string courses, arches (rubbed bricks) and special flashing details
- Eaves, parapets, copings and special mouldings

- Timber sections such as handrails, window sections and joinery details
- Jointing details of curtain walling and other specialised cladding
- Staircases
- Special joinery fixtures and fittings
- Any special feature which cannot be described or shown clearly to a smaller scale.

Schedules

It is good practice to convey information on items such as windows, doors and ironmongery by means of schedules which will set out the architect's requirements for other members of the building team. This can be done in a manner which has many advantages over drawings, for example:

- Checking for errors of omission or duplication is simplified.
- Quantifying items for the purpose of obtaining estimates or placing orders is simplified.
- If the information is set down systematically, prolonged searching through a specification is avoided.

In setting down information in schedules, ensure clear layout, simple coding, the minimum of abbreviation and proper reference to the location in the building of the items concerned. Schedules should be either the same size as the standard drawing sheets selected for the job or, if smaller, bound together as a set in one of the accepted paper sizes.

The design of and information contained in the title blocks should be consistent with those of the drawings. Schedules should also be numbered in a manner which is an extension of the numbering system used for the drawings. The first schedule should be a list of all the drawings and schedules related to the project, with space to record the latest revision number of each document. The first formal use of this schedule will be its inclusion in the specification or bills of quantities providing a record of the drawings and schedules used in the preparation of the tender documents. Schedules are not intended to supplant the specification or bills of quantities, which will contain the full and final description of the characteristics of the items concerned.

The layout of schedules will vary according to the information to be conveyed. Items being scheduled may be listed to read downwards on the left-hand side, with the characteristics and sizes of the items reading across or vice versa. Diagrams may well be included in the body of the schedule and a column reserved for notes and records of revisions. Once an item has been fully described in a schedule it should not be repeated at length. For example, having described the construction of a particular type of door, it can be given a type reference and only this would be repeated in the schedule. Characteristics, particularly on a finishes schedule, can often be defined by a code with an interpretation in the form of a key to one side. Constant repetition of descriptions is then eliminated. Well-recognised abbreviations which will not be confused with others can be used to save time.

Examples of schedules are given at the end of this chapter:

- Example 17.1 – Window schedule
- Example 17.2 – Door schedule
- Example 17.3 – Finishings schedule
- Example 17.4 – Manhole schedule.

This is by no means a complete and comprehensive list – there are many other items which may be described and collated conveniently in schedules, but it is very difficult for them to be standardised as they vary so much with each building. The examples given vary from those shown in BS EN ISO 4157:1999 but they are more easily understood and less likely to lead to error.

Drawings and schedules for records

Although we have already referred in this chapter to the keeping of record drawings, it is a matter of sufficient importance to justify emphasis and clarification. Before starting work on site there are two stages at which it is important to keep record drawings and schedules before they are subjected to any further amendment:

- Documents upon which the specification or bills of quantities are based – a full set of completed drawings and schedules.
- Contract drawings and schedules – those to which the contract documents refer and which are marked accordingly.

It is useful for these copies to be kept in such form as permits them to be reproduced should the need arise.

True and accurate 'as built' drawings, including the exact routes of all services, are vital in the future maintenance of the building and copies should be made and issued to the employer upon completion of the contract as part of the health and safety file. Their preparation depends upon regular correction of the drawings as the work proceeds and as amendments are issued. They may be in plastic negative form so that they are easily reproduced and read and microfilm used to minimise on storage space, although nowadays it is more likely that the drawings will be stored and made available electronically.

WINDOWS

WINDOW TYPE REF:	A	B	C	D	E
WINDOW REF NO.	W2 to W19 inc. W22 to W39 inc.	W1, W20 W21, W40	W41, W42 and WW3	W44 to W50 inc.	W51
NUMBER OFF	36	4	3	7	1
MATERIAL OF MAIN FRAME	Extruded aluminium PVF2 treated, insulated sections: by Glazing International Limited				
SUB-FRAME	Nil	Nil	Nil	Nil	Hardwood
BUILDERS OPENING SIZE	Continuous horizontal opening height 1200		600 H × 1500 W	1800 H × 750 W	1800 diam
WINDOW UNIT SIZE	1185 H × 1450 W (inc. cill)	1185 H × 750 W (inc. cill)	585 H × 1485 W	1785 H × 735 W	1485 diam
FIXING JAMB	To coupling mullion	Alum. brackets to brickwork	Aluminium brackets to brickwork		
HEAD	Alumium brackets plugged to concrete lintals or bean				Nil
CILL	Aluminium cill with brackets plugged and screwed to blockwork				Nil
SEALANT	Brown polysulphide to heads and jambs				
IRONMONGERY AND GLAZING					
FASTENINGS	Alum. lever handle and locking latch	Nil	Aluminium level handle	Aluminium peg stay	Nil
OPENING LIGHTS HINGES/PIVOTS	Horizontal hung projecting opening gear	Nil	Top hung (hinges with frame)	Side hung (hinges with frame)	Nil
STAYS	As part of opening gear	Nil	Pair of friction stays	Nil	Nil
OPERATING GEAR	Nil	Nil	Nil	Nil	Nil
SECURITY	Safety catch in jamb	Nil	Nil	Nil	Nil
GLAZING	Double glazed units, outer glass 6 mm Pickertons Antison Silver, 12 mm void, inner glass 6 mm clear float				
FIXING	Extruded neoprene gaskets – all windows pre glazed by manufacturer				
NOTES					See drawing No. Do2/64
REVISIONS Ref / Wind / Date A / WA/21 / 28/2/92	DIAGRAMS				

Project Title SHOPS & OFFICES – NEW BRIDGE STREET – BORCHESTER

Architects: REED & SEYMORE WINDOW SCHEDULE NO. 456/9.1 REV. A

Example 17.1 Window schedule.

DOORS

DOOR TYPE REF:	A		B	C	D
DOOR REF NO.	1	2 to 8	1 to 4	1 to 8	1 to 15
LOCATION AND NUMBER	Main entrance lobby 1	Entrance hall and corridor 01 & 11 7	Main staircase 4	Rooms 03, 04, 05 & 10 to 14 incl. 8	Offices 06 to 010 & 115 to 120 15
TYPE AND DESCRIPTION	HW framed single swing double door: 6 mm toughened glass		Plyfaced solid core flush panel with GWPP glazed viewing panel		Plyfaced solid core flush panel
FIRE RATING	Nil	Nil	60/60	30/30	Nil
DOOR SIZE	2 @ 2040 × 726 × 46 with 12 mm reboted styles		2040 × 826 × 46 and 2040 × 426 × 46	2040 × 826 × 46	
FINISH	Polyurathene sealed		Colour preservalive treated and wax polished		
FRAME OVERALL SIZE (NOMINAL) AND SECTION	2375 × 1850, 2375 × 1500 ex 150 × 60 rebated 25 mm kicking rail ex 200 × 36		2375 × 1350 ex 150 × 50 rebated 25 mm	2100 × 900 ex 150 × 60 rebated 25 mm	2100 × 900 125 × 32 plus stops
MATERIAL/ FINISH	Hardwood sealed		Softwood primed and painted		
FAN LIGHT	6 mm PP		6 mm GWPP	Nil	Nil
SIDE LIGHT	6 mm toughened glass	Nil	6 mm GWPP	Nil	Nil
SWING IRONMONGERY	1 @ 726 726	7 @ 726 726	4 @ 826 726	6 2	7 8
HINGES	1½ pair SS			1 pair rising butts	1 pair
LOCKS AND LATCHES	Mortice deadlock type Ref. type –	Mortice dead lock type –	Mortice dead lock type –	Mortice latch and dead lock type –	
HANDLES	1 pair 'D' pattern pull handles type –		Lever handles type –		
KICK PLATES	2 pairs satin aluminium 200 high		Nil	Nil	Nil
PUSH PLATES	1 pair satin aluminium 200 high		Nil	Nil	Nil
BOLTS	2 off SA flush (to leaf without lock)		2 off SA flush to small leaf	Nil	Nil
STOPS AND STAYS	Skirting mounted type –			Floor mounted door stop type –	
MISCELLANEOUS			Surface mounted overhead closers. Intumescent strip to 3 sides		
REVISIONS Ref A, Door DA/2, Date 28/2/92	DIAGRAMS				

Project Title SHOPS & OFFICES – NEW BRIDGE STREET – BORCHESTER

Architects: REED & SEYMORE DOOR SCHEDULE NO. 456/13.2 REV. A

Example 17.2 Door schedule.

FINISHES

FLOOR LEVEL	Ground	Ground	Ground	First	First
ROOM NAME	Entrance lobby	Stair lobby	Enquiries	General office	Womens lavatory
ROOM NO.	G01	G02	G03	101	102
FINISHES					
SCREED AND FLOOR FINISH	75 mm screed Carpel, Swatch ZD 10016			Raised floor system with carpet inlays trays	50 mm screed 150 × 150 ceramic floor tite
WALLS 1 2 3 4	Glazed Sirpite plaster finish to sand and cement base coat				Plaster Glazed wall tiles Glazed wall tiles Glazed wall tiles
SKIRTINGS	4 × 100 × 19 HW as detail 456/W/26			19 × 100 SW	Flush, coved ceramic tite
CEILING	Concealed grid suspension system; Class 1 tiles		Lay – in tile suspension system: Class 0/1		Plaster board, scrim and set
MISC AND NOTES	Standard brass mat well frame set into screed	Staircase see detail drawing 456/14.3	Reception counter see drawing 456/W/25		
DECORATIONS					
WALL FINISHES	Vinyl wall fabric by Sampsons Ltd	1 mist coat plus 2 coats vinyl emulsion – malt			Prime plus 2 coats flat oil
COLOURS 1 2 3 4	Glazed Ref No. SAM 1234 Ref No. SAM 1234 Ref No. SAM 1234	BS 00A01 BS 00A01 BS 00A01 BS 00A01	BS 20C37 BS 20C33 BS 20C33 BS 20C33	BS 16C33 BS 16C33 BS 16C33 BS 16C33	BS 18E51 2, 3 & 4 tiled
CEILINGS	Nil	Nil	Nil	Nil	Flat oil white
FRAMES AND ARCHITRAVES	1 + 2 coat gloss oil brilliant white				H2 gloss oil BS 20C40
DOORS	Glazed alumin	HW veneer 2 coats satin polyeurathene seal			Plaslic laminate faced
WINDOWS	Aluminium frames self finished throughout				
CILLS	1 + 2 coats gloss oil brilliant white				Tiled as walls
SKIRTINGS	1 + 2 coats gloss oil brilliant white				1 + 2 gloss oil BS 20C40
REVISIONS Ref / Rm No. / Date A / C01 / 1/3/92 B / I02 / 16/3/92	MISC AND NOTES	Handrails – 2 coats glass polyeurathen seal	Reception counter – American oak french polished		Plaslic laminate cubicles and ducting, extent of wall tiling etc. see drawing 456/W/27

Project Title SHOPS & OFFICES – NEW BRIDGE STREET – BORCHESTER

Architects: REED & SEYMORE FINISHINGS SCHEDULE NO. 456/14.1 REV. B

Example 17.3 Finishings schedule.

MANHOLES AND COVERS

MANHOLE REF:	FM1	FM2	FM3	FM4	FM5
INIERNAL SIZE	1125 × 825	900 diam	1050 diam	1050 diam	1350 reducing to 900 diam
CONSTRUCTION AND MAKE	225 mm brick to 150 conc base	Precost concrete chamber rings, slabs and seating rings set in concrete backlill. 150 mm insitu concrete base			
COVER SIZE	450 × 600	450 × 600			
COVER TYPE	Med duty double – sealed	Grade C Fig 7		Grade C Fig 5	
INVERT LEVEL	7.315	7.214	7.02	6.807	5.705
GROUND LEVEL	8.46	8.153	8.175	8.08	8.07
COVER LEVEL	8.46	8.382	8.175	8.382	8.382
MAIN CHANNEL SIZE	100 mm	100 mm	100 mm	100/150 mm	150 mm
TYPE STRAIGHT (S) CURVED (C) TAPPERED (T)	S	C	S	S T	S
BRANCH CHANNELS SIZE TO RH	2 × 100 mm × 90°		1 × 100 mm × 90° 1 × 100 mm × 135°	1 × 150 mm × 90°	
SIZE TO LH	1 × 100 mm × 135°	1 × 100 mm × 90°	1 × 100 mm × 135°	1 × 100 mm × 90° 1 × 100 mm × 135°	1 × 100 mm × 90°
INTERCEPTING TRAP	100 mm				
F.A.I.					100 mm
BACKDROP INLET					As detail drawing 071/C
STEP IRONS					C.I. into precast units
MISCELLANEOUS	OPC between head of brick and G.F stab		MH cover roised to paving level on brick upstand		
REVISIONS Ref · MH No · Date A · FM4 · 28/2/92	DIAGRAMS 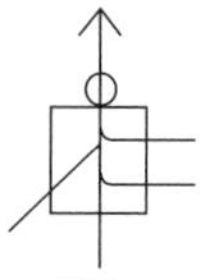 Internal MH	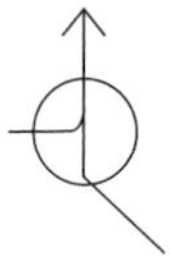			

Project Title SHOPS & OFFICES – NEW BRIDGE STREET – BORCHESTER

Architects: REED & SEYMORE MANHOLES & COVER SCHEDULE NO. 456/ REV. A

Example 17.4 Manhole schedule.

Chapter 18
Specifications

The use of specifications

The types and uses of specifications have changed dramatically in the past decade. They are no longer seen as documents that supplement the drawings or bills of quantities but as the key, stand-alone documents that define the scope and quality requirements in any construction contract.

For many years specifications comprised preambles to the bills of quantities, or brief written statements clarifying issues which could not be shown on the drawings. Occasionally they were used to describe a sequence of tasks to be performed. Today the picture is very different. Preambles have disappeared, drawings are produced using CAD technology and bills of quantities are often no more than a pricing schedule. The specification acts as the link between the matrices of the contract documentation, detailing not only scope, materials and workmanship requirements but also visual, procedural, design, testing, quality control and responsibility issues – the complete picture. The specification is now considered a prime document in any construction contract, second only in importance to the terms and conditions, for the simple reason that it records 'the buy'.

In today's complex array of procurement, design and contractual methods, which seek to utilise the expertise of many contributors, the specification is the document that confirms precisely what one party agrees to provide to the other in respect of scope and quality. An inadequate clause or requirement, which conflicts with other contractual statements, will almost certainly lead to unwanted additional costs. The construction industry no longer relies on the 'design it, buy it then build it' methodology. Employer demands for high quality, speed of construction and cost control increasingly lead to the need for off-site fabrication where factory environments facilitate better quality control, economies of scale and less time on site. Traditional construction documentation is not always relevant to or understood by the manufacturing industry, and specifications therefore have to be adapted to suit. Modern buildings are very much a team effort but legal responsibilities demand clear confirmations of who does what – the specification fulfils that requirement.

There are three basic types of specification – prescriptive, performance, or descriptive.

Prescriptive specifications

More commonly known as a 'detailed materials and workmanship specification', this is a 'do precisely as I say' specification. Such an approach pre-supposes that the design team know exactly what they want, have worked out exactly how it should be done, are allowed by European legislation to specify products, manufacturers, etc. and are prepared to warrant the design, taking full responsibility for the end product being fit for purpose. Prescriptive specifications are most commonly written for smaller projects using well-tried and tested technology.

Performance specifications

These are used in circumstances where the design team have absolutely no visual requirements but wish the contractor to select suitable materials and install them to meet stated criteria. A performance specification does not confirm what will be provided, only what has to be achieved, and as such employers do not know what they are getting when entering into a contract. The employer's requirement is only that

...for smaller projects using well-tried and tested technology

the end product performs and meets legislation and national standard requirements – it is a 'give me something that works' specification.

Descriptive specifications

The 'new boys on the block' specification, this hybrid has been developed to make best use of the varying skills of those involved in the process of delivering modern buildings. The main function of a descriptive specification is to define scope, design intent, procedures for completing detailed design, quality control and to provide the contractor with a fair indication of the solutions that are acceptable. Contractors are required to use their specialist expertise to complete the detailed design (in consultation with the design team), manufacture and install the works, and provide the necessary warranties and guarantees.

The need to prepare a contract specification which fully represents 'the buy' is fundamental to the successful use of a descriptive specification if claims and variations are to be avoided at a later date. This requires a process of evaluating tenders in order to agree the key elements that reflect the scope of works. Descriptive specifications require that the design team and the contractor work in harmony and adopt a 'help me find the best solution' approach.

Large complex buildings tend to require the descriptive approach while the more traditional design solutions lend themselves to prescriptive documents.

Specification writing

Having established the various types and uses of specifications, there are a number of ways in which they can be prepared. Generally speaking, the following five processes are necessary in the production of a good specification – decide on format, collect information, input information, check and test, and deliver.

Decide on format

When selecting a format, three options are available – uniclass, masterformat, or bespoke.

Uniclass

The Unified Classification for the Construction Industry published by RIBA Publications is a classification scheme for organising library materials and for structuring product literature and project information. Section J, which comprises work sections for buildings, is reproduced at the end of this chapter as Figure 18.1 with the permission of RIBA Enterprises Ltd. Uniclass can also be used by the architect for the classification of drawings (see Chapter 17) and by the quantity surveyor, via SMM7R, in the preparation of bills of quantities (see Chapter 19). It therefore provides useful links between the drawings, the bills of quantities (the pricing document) and the specification (the scope and quality document).

Care is required when using section JA because it mixes specification requirements for the permanent works with preliminaries requirements necessary to allow tenderers to price temporary, non-measurable and time-related items. It is important to remember that in the UK (unlike the USA) the specification is rarely the pricing document and bills of quantities no longer serve as the specification. Whilst the two should be cross-referenced for ease of use they should not be intermixed, although the specification may be incorporated in the bills of quantities in order to give it contract document status (see later in this chapter).

Another point worth noting is that the Uniclass work sections for buildings do not reflect construction trades or work packages. This results in most specifications being made up of several sections forming a rather bulky document. However, the proper use of sections JA7 (General specification for work packages) and JZ (Building fabric reference specification) helps to reduce repetition.

Masterformat

The Masterformat classification system is produced by the Construction Specifications Institute in the USA. It is the system that is recognised worldwide and is best adopted when working outside the UK or when working with large American firms in the UK. Masterformat is more construction trade related than Uniclass but it does include Section 1 (General Condition) which is similar to preliminaries in the UK. It is worth noting that bills of quantities are not generally recognised in the USA and that the specification is the document used for pricing.

Bespoke

Major employers commissioning multiple buildings occasionally produce specifications tailored to suit their particular needs, based on building type/size or a partnering procurement strategy. In most situations, however, the bespoke approach is not recommended as it is preferable to follow a recognised published format.

Collect information

It is to be remembered that the main purpose of any specification is to convey information from one party to another for contractual and control purposes.

Accurate information is essential in order to provide 'best information'. This can take many forms, ranging from the stipulation of a precise material and manufacturer to a detailed description of design intent. Both are equally effective provided that the specification format suits the procurement method and the selected form of contract. The wrong level of information put into a specification can lead to variations and claims for loss and/or expense.

One good method of collection is to produce a list of materials or system descriptions, each with a unique short code such as 'BLK-1 = 100 mm common blockwork'. These codes can be used on the drawings and in the specification to provide clear scope and definition. This system is commonly known as technical sheet collection.

The design team should decide at an early stage how best to gather information, bearing in mind the time available. Start collection early. Information should be updated continuously until the specification issue date. Continuous drafting and updating in parallel with drawing production is the best methodology. All changes should be recorded for quality control purposes.

Input information

The key to accurate and fast completion of a specification is to use reliable and up-to-date source data. Off-the-shelf products are available. The best for use in the UK is perhaps the National Building Specification (NBS) which is arranged under the Uniclass work sections for buildings. It comprises detailed specification clauses together with skeleton clauses providing a sound checklist of the information necessary to specify properly preliminaries, general conditions, materials, goods and workmanship. NBS is updated regularly to take account of amendments to the JCT forms and to reflect changes in technology and the latest standards for materials, goods and workmanship.

Masterspec is perhaps the best product for use in the USA when adopting Masterformat. Web-based products are also becoming available which usually provide up-to-date information and a fast production system. Whatever the source data, it is good practice for every clause in the specification to be uniquely numbered and for the project title, work section name and/or reference, version number, issue date and author's name to be given on every page. It is also good practice for drafts of the specification to be issued regularly for review and co-ordination by the design team.

Check and test

It is essential to ensure that there will be sufficient time available in which to test and check the specification before delivery. The author of a specification will never actually know how good it is until it is in use as a working document and is shown to provide clear direction when problems are encountered on site. However, it is a worthwhile exercise to test all specifications prior to finalising them to ensure that they include the appropriate clauses adequately to resolve all problems previously encountered in work of a similar nature.

As for checking, it is almost impossible for the same person to be able to write and accurately proofread a document because authors rarely spots their own mistakes. It is therefore essential to have a specification read through by a third party, who is preferably himself a competent specification writer and able to spot not only typographical errors but also where a mistake has been made or something does not make sense.

Deliver

There are two ways to deliver (issue) a specification – hard copy or electronic copy.

Hard copy

This involves numerous paper copies being delivered to all users and interested parties. These copies require close control and a record must be kept of exactly which version has been issued to whom, when and for what purpose. This method of delivery is expensive, slow, cumbersome and uses vast quantities of paper.

Electronic copy

The construction industry is slowly adapting to technology, and the use of electronic document management systems and web-based sites as a means of recording, storing and circulating information is on the increase. The use of CD-ROMs to circulate documents gives rise to similar problems as the use of hard copy, except there is less bulk and less paper used. The use of e-mail is preferable as it is fast, reliable and cheap. However, document identification and circulation control is just as important as it is with hard copy. Also, special consideration should be given to format – each recipient needs to be using the same software and settings in order to ensure that what everyone actually sees is the same document. It is strongly recommended that electronic files be issued always in a protected unexecutable format, such as PDF, in order to ensure that recipients cannot readily make inappropriate alterations – when that happens it can be disastrous. If any amendments to the specification are required for any reason, these should be raised with the design team and the specification ought only to be altered with express instruction.

Web-based, dynamic production systems are becoming available, providing faster, cheaper and better quality methods of production.

The specification as a contract document

Whilst the SBC Without Quantities recognises the specification as a contract document, the SBC With Quantities does not. When using the latter form of contract, the specification can be given contract document status either by amending the wording of the standard form (not to be recommended) or by incorporating the specification into the bills of quantities, which are recognised as part of the contract documentation.

Work sections for buildings J

JA Preliminaries/General conditions

JA1 The project generally
- JA10 Project particulars
- JA11 Documentation
- JA12 The site/Existing buildings
- JA13 Description of the work

JA2 The Contract
- JA20 The Contract/Sub-contract

JA3 Employer's requirements
- JA30 Tendering/Sub-letting/Supply
- JA31 Provision, content and use of documents
- JA32 Management of the Works
- JA33 Quality standards/control
- JA34 Security/Safety/Protection
- JA35 Specific limitations on method/sequence/timing/use of site
- JA36 Facilities/Temporary works/Services
- JA37 Operation/Maintenance of the finished building

JA4 Contractor's general cost items
- JA40 Management and staff
- JA41 Site accommodation
- JA42 Services and facilities
- JA43 Mechanical plant
- JA44 Temporary works

JA5 Work by others or subject to instruction
- JA50 Work/Materials by the employer
- JA51 Nominated sub-contractors
- JA52 Nominated suppliers
- JA53 Work by statutory authorities
- JA54 Provisional work
- JA55 Dayworks

JA6 Preliminaries for specialist contracts
- JA60 Demolition contract preliminaries
- JA61 Ground investigation contract preliminaries
- JA62 Piling contract preliminaries
- JA63 Landscape contract preliminaries

JA7 General specification for work packages
- JA70 General specification for building fabric work
- JA71 General specification for building services work

JB Complete buildings/structures/units

JB1 Prefabricated buildings/structures/units
- JB10 Prefabricated buildings/structures
- JB11 Prefabricated building units

JC Existing site/buildings/services

JC1 Investigations/Surveys
- JC10 Site survey
- JC11 Ground investigation
- JC12 Underground services survey
- JC13 Building fabric survey
- JC14 Building services survey

JC2 Demolition/Removal
- JC20 Demolition
- JC21 Toxic/hazardous material removal

JC3 Alteration – support
- JC30 Shoring/Facade retention

JC4 Repairing/Renovating/Conserving concrete/masonry
- JC40 Cleaning masonry/concrete
- JC41 Repairing/Renovating/Conserving masonry
- JC42 Repairing/Renovating/Conserving concrete
- JC45 Damp proof course renewal/insertion

JC5 Repairing/Renovating/Conserving metal/timber
- JC50 Repairing/Renovating/Conserving metal
- JC51 Repairing/Renovating/Conserving timber
- JC52 Fungus/Beetle eradication

JC9 Alteration – composite items
- JC90 Alterations – spot items

JD Groundwork

JD1 Ground stabilisation/dewatering
- JD11 Soil stabilisation
- JD12 Site dewatering

JD2 Excavation/filling
- JD20 Excavating and filling
- JD21 Landfill capping

JD3 Piling
- JD30 Piling

JD4 Ground retention
- JD40 Embedded retaining walls
- JD41 Crib walls/Gabions/Reinforced earth

JD5 Underpinning
- JD50 Underpinning

JE In situ concrete/Large precast concrete

JE0 Concrete construction generally
- JE05 In situ concrete construction generally

JE1 Mixing/Casting/Curing/Spraying in situ concrete
- JE10 Mixing/Casting/Curing in situ concrete
- JE11 Sprayed concrete

JE2 Formwork
- JE20 Formwork for in situ concrete

JE3 Reinforcement
- JE30 Reinforcement for in situ concrete
- JE31 Post tensioned reinforcement for in situ concrete

JE4 In situ concrete sundries
- JE40 Designed joints in in situ concrete
- JE41 Worked finishes/Cutting to in situ concrete
- JE42 Accessories cast into in situ concrete

JE5 Structural precast concrete
- JE50 Precast concrete frame structures

JE6 Composite construction
- JE60 Precast/Composite concrete decking

Figure 18.1 Uniclass Section J (RIBA Enterprises).

J Work sections for buildings

JF Masonry

JF1 Brick/Block walling
- JF10 Brick/Block walling
- JF11 Glass block walling

JF2 Stone walling
- JF20 Natural stone rubble walling
- JF21 Natural stone ashlar walling/dressings
- JF22 Cast stone walling/dressings

JF3 Masonry accessories
- JF30 Accessories/Sundry items for brick/block/stone walling
- JF31 Precast concrete sills/lintels/copings/features

JG Structural/Carcassing metal/timber

JG1 Structural/Carcassing metal
- JG10 Structural steel framing
- JG11 Structural aluminium framing
- JG12 Isolated structural metal members

JG2 Structural/Carcassing timber
- JG20 Carpentry/Timber framing/First fixing

JG3 Metal/Timber decking
- JG30 Metal profiled sheet decking
- JG31 Prefabricated timber unit decking
- JG32 Edge supported/Reinforced woodwool slab decking

JH Cladding/Covering

JH1 Glazed cladding/covering
- JH10 Patent glazing
- JH11 Curtain walling
- JH12 Plastics glazed vaulting/walling
- JH13 Structural glass assemblies
- JH14 Concrete rooflights/pavement lights
- JH15 Rainscreen cladding/overcladding

JH2 Sheet/board cladding
- JH20 Rigid sheet cladding
- JH21 Timber weatherboarding

JH3 Profiled/flat sheet cladding/covering
- JH30 Fibre cement profiled sheet cladding/covering
- JH31 Metal profiled/flat sheet cladding/covering
- JH32 Plastics profiled sheet cladding/covering
- JH33 Bitumen and fibre profiled sheet cladding/covering

JH4 Panel cladding
- JH40 Glass reinforced cement panel cladding/features
- JH41 Glass reinforced plastics panel cladding/features
- JH42 Precast concrete panel cladding/features
- JH43 Metal panel cladding/features

JH5 Slab cladding
- JH50 Precast concrete slab cladding/features
- JH51 Natural stone slab cladding/features
- JH52 Cast stone slab cladding/features

JH6 Slate/Tile cladding/covering
- JH60 Plain roof tiling
- JH61 Fibre cement slating
- JH62 Natural slating
- JH63 Reconstructed stone slating/tiling
- JH64 Timber shingling
- JH65 Single lap roof tiling
- JH66 Bituminous felt shingling

JH7 Malleable sheet coverings/cladding
- JH70 Malleable metal sheet pre-bonded coverings/claddings
- JH71 Lead sheet coverings/flashings
- JH72 Aluminium sheet coverings/flashings
- JH73 Copper strip/sheet coverings/flashings
- JH74 Zinc strip/sheet coverings/flashings
- JH75 Stainless steel strip/sheet coverings/flashings
- JH76 Fibre bitumen thermoplastic sheet coverings/flashings

JH9 Other cladding/covering
- JH90 Tensile fabric coverings
- JH91 Thatch roofing

JJ Waterproofing

JJ1 Cementitious coatings
- JJ10 Specialist waterproof rendering

JJ2 Asphalt coatings
- JJ20 Mastic asphalt tanking/damp proofing
- JJ21 Mastic asphalt roofing/insulation/finishes
- JJ22 Proprietary roof decking with asphalt finish

JJ3 Liquid applied coatings
- JJ30 Liquid applied tanking/damp proofing
- JJ31 Liquid applied waterproof roof coatings
- JJ32 Sprayed vapour barriers
- JJ33 In situ glass reinforced plastics

JJ4 Felt/flexible sheets
- JJ40 Flexible sheet tanking/damp proofing
- JJ41 Built-up felt roof coverings
- JJ42 Single layer polymeric roof coverings
- JJ43 Proprietary roof decking with felt finish
- JJ44 Sheet linings for pools/lakes/waterways

JK Linings/Sheathing/Dry partitioning

JK1 Rigid sheet sheathing/linings
- JK10 Plasterboard dry lining/partitions/ceilings *For panel partitions see JK30.*
- JK11 Rigid sheet flooring/sheathing/linings/casings
- JK12 Under purlin/Inside rail panel linings
- JK13 Rigid sheet fine linings/panelling
- JK14 Glass reinforced gypsum linings/panelling/casings/mouldings
- JK15 Vitreous enamel linings/panelling

JK2 Timber board/Strip linings
- JK20 Timber board flooring/sheathing/linings/casings
- JK21 Timber strip/board fine flooring/linings

Figure 18.1 (*Cont.*)

Work sections for buildings J

JK3 Dry partitions
- JK30 Panel partitions
- JK31 *intentionally not used*
- JK32 Framed panel cubicles
- JK33 Concrete/Terrazzo partitions

JK4 False ceilings/floors
- JK40 Demountable suspended ceilings
- JK41 Raised access floors

JL Windows/Doors/Stairs

JL1 Windows/Rooflights/ Screens/Louvres
- JL10 Windows
- JL11 Rooflights/ Roof windows
- JL12 Screens
- JL13 Louvred ventilators
- JL14 External louvres/ shutters/canopies/blinds

JL2 Doors/Shutters/Hatches
- JL20 Doors
- JL21 Shutters
- JL22 Hatches

JL3 Stairs/Walkways/Balustrades
- JL30 Stairs/Walkways/ Balustrades

JL4 Glazing
- JL40 General glazing
- JL41 Lead light glazing
- JL42 Infill panels/sheets

JM Surface finishes

JM1 Screeds/Trowelled flooring
- JM10 Cement:sand/Concrete screeds/toppings
- JM11 Mastic asphalt flooring/ floor underlays
- JM12 Trowelled bitumen/resin/ rubber-latex flooring
- JM13 Calcium sulfate based screeds

JM2 Plastered coatings
- JM20 Plastered/Rendered/ Roughcast coatings
- JM21 Insulation with rendered finish
- JM22 Sprayed monolithic coatings
- JM23 Resin bound mineral coatings

JM3 Work related to plastered coatings
- JM30 Metal mesh lathing/ Anchored reinforcement for plastered coatings
- JM31 Fibrous plaster

JM4 Rigid tiles
- JM40 Stone/Concrete/Quarry/ Ceramic tiling/Mosaic
- JM41 Terrazzo tiling/In situ terrazzo
- JM42 Wood block/Composition block/Parquet flooring

JM5 Flexible sheet/tile coverings
- JM50 Rubber/Plastics/Cork/ Lino/Carpet tiling/ sheeting
- JM51 Edge fixed carpeting
- JM52 Decorative papers/fabrics

JM6 Painting
- JM60 Painting/Clear finishing
- JM61 Intumescent coatings for fire protection of steelwork

JN Furniture/Equipment

JN1 General purpose fixtures/ furnishings/equipment
- JN10 General fixtures/ furnishings/equipment
- JN11 Domestic kitchen fittings
- JN12 Catering equipment
- JN13 Sanitary appliances/ fittings
- JN14 Plant containers
- JN15 Signs/Notices

JN2 Special purpose fixtures/ furnishings/equipment
- JN20 Appropriate section title for each project
- JN21 Appropriate section title for each project
- JN22 Appropriate section title for each project
- JN23 Appropriate section title for each project

JP Building fabric sundries

JP1 Sundry proofing/insulation
- JP10 Sundry insulation/ proofing work/fire stops
- JP11 Foamed/Fibre/Bead cavity wall insulation

JP2 Sundry finishes/fittings
- JP20 Unframed isolated trims/skirtings/sundry items
- JP21 Ironmongery
- JP22 Sealant joints

JP3 Sundry work in connection with engineering services
- JP30 Trenches/Pipeways/Pits for buried engineering services
- JP31 Holes/Chases/Covers/ Supports for services

JQ Paving/Planting/Fencing/Site furniture

JQ1 Edgings/Accessories for pavings
- JQ10 Kerbs/Edgings/Channels/ Paving accessories

JQ2 Pavings
- JQ20 Granular sub-bases to roads/pavings
- JQ21 In situ concrete roads/pavings/bases
- JQ22 Coated macadam/ Asphalt roads/pavings
- JQ23 Gravel/Hoggin/Bark roads/pavings
- JQ24 Interlocking brick/block roads/pavings
- JQ25 Slab/Brick/Sett/Cobble pavings
- JQ26 Special surfacings/ pavings for sport/general amenity

JQ3 Planting
- JQ30 Seeding/Turfing
- JQ31 Planting
- JQ32 Planting in special environments
- JQ35 Landscape maintenance

JQ4 Fencing
- JQ40 Fencing

JQ5 Site furniture
- JQ50 Site/Street furniture/ equipment

Figure 18.1 (*Cont.*)

J Work sections for buildings

JR Disposal systems

JR1 Drainage
- JR10 Rainwater pipework/gutters
- JR11 Foul drainage above ground
- JR12 Drainage below ground
- JR13 Land drainage
- JR14 Laboratory/Industrial waste drainage

JR2 Sewerage
- JR20 Sewage pumping
- JR21 Sewage treatment/sterilisation

JR3 Refuse disposal
- JR30 Centralised vacuum cleaning
- JR31 Refuse chutes
- JR32 Compactors/Macerators
- JR33 Incineration plant

JS Piped supply systems

JS1 Water supply
- JS10 Cold water
- JS11 Hot water
- JS12 Hot and cold water (small scale)
- JS13 Pressurised water
- JS14 Irrigation
- JS15 Fountains/Water features

JS2 Treated on site water supply
- JS20 Treated/Deionised/Distilled water
- JS21 Swimming pool water treatment

JS3 Gas supply
- JS30 Compressed air
- JS31 Instrument air
- JS32 Natural gas
- JS33 Liquefied petroleum gas
- JS34 Medical/Laboratory gas

JS4 Petrol/Oil storage
- JS40 Petrol/Diesel storage/distribution
- JS41 Fuel oil storage/distribution

JS5 Other supply systems
- JS50 Vacuum
- JS51 Steam

JS6 Fire fighting – water
- JS60 Fire hose reels
- JS61 Dry risers
- JS62 Wet risers
- JS63 Sprinklers
- JS64 Deluge
- JS65 Fire hydrants

JS7 Fire fighting – gas/foam
- JS70 Gas fire fighting
- JS71 Foam fire fighting

JT Mechanical heating/Cooling/Refrigeration systems

JT1 Heat source
- JT10 Gas/Oil fired boilers
- JT11 Coal fired boilers
- JT12 Electrode/Direct electric boilers
- JT13 Packaged steam generators
- JT14 Heat pumps
- JT15 Solar collectors
- JT16 Alternative fuel boilers

JT2 Primary heat distribution
- JT20 Primary heat distribution

JT3 Heat distribution/utilisation – water
- JT30 Medium temperature hot water heating
- JT31 Low temperature hot water heating
- JT32 Low temperature hot water heating (small scale)
- JT33 Steam heating

JT4 Heat distribution/utilisation – air
- JT40 Warm air heating
- JT41 Warm air heating (small scale)
- JT42 Local heating units

JT5 Heat recovery
- JT50 Heat recovery

JT6 Central refrigeration/Distribution
- JT60 Central refrigeration plant
- JT61 Chilled water

JT7 Local cooling/Refrigeration
- JT70 Local cooling units
- JT71 Cold rooms
- JT72 Ice pads

JU Ventilation/Air conditioning systems

JU1 Ventilation/Fume extract
- JU10 General ventilation
- JU11 Toilet ventilation
- JU12 Kitchen ventilation
- JU13 Car parking ventilation
- JU14 Smoke extract/Smoke control
- JU15 Safety cabinet/Fume cupboard extract
- JU16 Fume extract
- JU17 Anaesthetic gas extract

JU2 Industrial extract
- JU20 Dust collection

JU3 Air conditioning – all air
- JU30 Low velocity air conditioning
- JU31 VAV air conditioning
- JU32 Dual-duct air conditioning
- JU33 Multi-zone air conditioning

JU4 Air conditioning – air/water
- JU40 Induction air conditioning
- JU41 Fan-coil air conditioning
- JU42 Terminal re-heat air conditioning
- JU43 Terminal heat pump air conditioning

JU5 Air conditioning – hybrid
- JU50 Hybrid system air conditioning

JU6 Air conditioning – local
- JU60 Air conditioning units

JU7 Other air systems
- JU70 Air curtains

JV Electrical supply/power/lighting systems

JV1 Generation/Supply/HV distribution
- JV10 Electricity generation plant
- JV11 HV supply/distribution/public utility supply
- JV12 LV supply/public utility supply

JV2 General LV distribution/lighting/power
- JV20 LV distribution
- JV21 General lighting
- JV22 General LV power

JV3 Special types of supply/distribution
- JV30 Extra low voltage supply
- JV31 DC supply
- JV32 Uninterrupted power supply

Figure 18.1 *(Cont.)*

Work sections for buildings J

JV4 Special lighting
JV40 Emergency lighting
JV41 Street/Area/Flood lighting
JV42 Studio/Auditorium/Arena lighting

JV5 Electric heating
JV50 Electric underfloor/ceiling heating
JV51 Local electric heating units

JV9 General/Other electrical work
JV90 General lighting and power (small scale)

JW Communications/Security/Control systems

JW1 Communications – speech/audio
JW10 Telecommunications
JW11 Paging/Emergency call
JW12 Public address/Conference audio facilities

JW2 Communications – audio-visual
JW20 Radio/TV/CCTV
JW21 Projection
JW22 Information/Advertising display
JW23 Clocks

JW3 Communications – data
JW30 Data transmission

JW4 Security
JW40 Access control
JW41 Security detection and alarm

JW5 Protection
JW50 Fire detection and alarm
JW51 Earthing and bonding
JW52 Lightning protection
JW53 Electromagnetic screening
JW54 Liquid detection alarm
JW55 Gas detection alarm
JW56 Electronic bird/vermin control

JW6 Central control
JW60 Central control/Building management

JX Transport systems

JX1 People/Goods
JX10 Lifts
JX11 Escalators
JX12 Moving pavements
JX13 Powered stairlifts
JX14 Fire escape chutes/slings

JX2 Goods/Maintenance
JX20 Hoists
JX21 Cranes
JX22 Travelling cradles/Gantries/Ladders
JX23 Goods distribution/Mechanised warehousing

JX3 Documents
JX30 Mechanical document conveying
JX31 Pneumatic document conveying
JX32 Automatic document filing and retrieval

JY Services reference specification

JY1 Pipelines and ancillaries
JY10 Pipelines
JY11 Pipeline ancillaries

JY2 General pipeline equipment
JY20 Pumps
JY21 Water tanks/cisterns
JY22 Heat exchangers
JY23 Storage cylinders/Calorifiers
JY24 Trace heating
JY25 Cleaning and chemical treatment

JY3 Air ductlines and ancillaries
JY30 Air ductlines/ancillaries

JY4 General air ductline equipment
JY40 Air handling units
JY41 Fans
JY42 Air filtration
JY43 Heating/Cooling coils
JY44 Air treatment
JY45 Silencers/Acoustic treatment
JY46 Grilles/Diffusers/Louvres

JY5 Other common mechanical items
JY50 Thermal insulation
JY51 Testing and commissioning of mechanical services
JY52 Vibration isolation mountings
JY53 Control components – mechanical
JY54 Identification – mechanical
JY59 Sundry common mechanical items

JY6 Cables and wiring
JY60 Conduit and cable trunking
JY61 HV/LV cables and wiring
JY62 Busbar trunking
JY63 Support components – cables

JY7 General electrical equipment
JY70 HV switchgear
JY71 LV switchgear and distribution boards
JY72 Contactors and starters
JY73 Luminaires and lamps
JY74 Accessories for electrical services

JY8 Other common electrical items
JY80 Earthing and bonding components
JY81 Testing and commissioning of electrical services
JY82 Identification – electrical
JY89 Sundry common electrical items

JY9 Other common mechanical and/or electrical items
JY90 Fixing to building fabric
JY91 Off-site painting/Anti-corrosion treatments
JY92 Motor drives – electric

JZ Building fabric reference specification

JZ1 Fabricating
JZ10 Purpose made joinery
JZ11 Purpose made metalwork
JZ12 Preservative/Fire retardant treatments for timber

JZ2 Fixing/Jointing
JZ20 Fixings/Adhesives
JZ21 Mortars
JZ22 Sealants

JZ3 Finishing
JZ30 Off-site painting
JZ31 Powder coatings
JZ32 Liquid coatings
JZ33 Anodising

Figure 18.1 *(Cont.)*

Chapter 19
Bills of Quantities

Tender and contract document

Bills of quantities are documents that describe the quality and give the quantities of the constituent parts of proposed building works. They have two primary functions. Initially they are used as tender procurement documents to provide a uniform basis for competitive lump sum tenders. Subsequently they become contract documents serving as schedules of rates for the pricing of variations. It is therefore important that they contain at least the basic information required of them by the prevailing

Presented in a recognisable format

conditions of contract and that they are presented in a recognisable format that will facilitate their use.

The SBC With Quantities requires bills of quantities to be prepared in accordance with the current version of the Standard Method of Measurement of Building Works 7th Edition (SMM7R at the time of publication).

The wider role

Even with the more or less universally adopted use of computers, the cost of producing bills of quantities is high and consequently their use as tender procurement and contract documents is often questioned and new techniques and procedures aimed at supplanting them are often tried. However, bills of quantities have survived thus far, no doubt because in addition to their two primary functions they contain vast amounts of information that can be of use in many ways. Also, they are frequently used in the preparation of interim valuations (see Chapter 28).

In preparing bills of quantities the quantity surveyor sifts through and processes much detailed information that, although not required by SMM7R to be included in the bills of quantities, may be of use to tendering contractors and the building team in connection with project planning, administration and the compilation of historical cost data in the form of elemental cost analyses. It is therefore worthwhile, in the early stages of the plan of work, to consider what additional information can be readily gathered in the measurement process and/or presented in the bills of quantities to assist tenderers and the building team to operate more effectively and efficiently.

Basic information

Bills of quantities are prepared by describing and quantifying, in accordance with the rules of SMM7R, the work shown on the drawings and given in the specification. Amongst other things SMM7R requires that 'Bills of Quantities shall fully describe and accurately represent the quantity and quality of the works to be carried out.'

It is normal practice to arrange bills of quantities in three generic sections that between them 'define the precise nature and extent of the required work' (SMM7R General Rule 1.1). These are preliminaries, preambles and measured works.

Preliminaries

The preliminaries are normally arranged in the Uniclass preliminaries/general conditions sections (see Chapter 18). The aim is to provide a comprehensive schedule of items relating to the project generally, the contract, the 'employer's requirements', the contractor's general cost items and work by others that can be individually priced by contractors for the purposes of calculating their lump sum price and for fixing rates that can be subsequently used in the valuation of variations. Within this framework it is necessary to provide at least the following information:

- name, nature and location of the project
- names and addresses of the employer and consultants
- list of drawings and other tender and contract documents
- site boundaries, existing buildings and existing mains services
- description of the works, including the dimensions and shape of each building
- form of contract, including a schedule of standard clause headings, details of any amendments to standard conditions and insertions into the contract particulars
- 'employer's requirements' and any limitations imposed by the employer relating to tendering, sub-letting, supply, provision, content and use of documents, management of the works, quality standards/control, security, safety, protection, method of working, sequence of work, timing, use of site, temporary facilities, temporary works, temporary services and operation/maintenance of the finished building
- contractor's general cost items relating to management, staff, site accommodation, services, facilities, mechanical plant and temporary works
- work or products by or from others directly employed by the employer
- provisional sums for work by local authorities and statutory undertakers
- provisional sums for the labour, materials and plant elements of dayworks
- provisional sums for defined and undefined provisional work.

Provisional sums are defined in both the SBC and SMM7R. They are used to make cost provision within the bills of quantities for works to be carried out by local authorities or statutory undertakers, dayworks and other defined and undefined provisional work that cannot be described and given in items in accordance with the rules of SMM7R. Also the SBC requires the architect contract administrator to issue instructions in regard to the expenditure of provisional sums.

General Rule 10 of SMM7R clearly sets down what is meant by defined and undefined provisional work and this rule is referred to in the definitions section of the SBC with quantities. Provided all of the required information is given in the description of defined provisional work attached to a provisional sum, the contractor is deemed to have made due allowance in his tender for that work in programming, planning and pricing preliminaries. However, where all of the required information is not given, the work is deemed to be undefined provisional work and the contractor will be deemed not to have made any such allowance. It is therefore of utmost importance that provisional sums for undefined provisional work include due allowance to cover the costs deemed not to have been included by the contractor in his lump sum price. It is also worthy of note that, when the naming of a supplier or subcontractor or work by local authorities or statutory undertakers occurs as a result of expenditure of a provisional sum, that sum will have to be sufficient to cover also the cost of the main contractor's profit and attendances, and the fixing only of any supplied items.

Preambles

The preambles traditionally comprise clauses specifying the quality or standard of materials and workmanship required. However, the preambles now more commonly comprise a simple reference to a separate, comprehensive specification document (see Chapter 18) which is thereby incorporated into the bills of quantities and given contract document status. Either way these specification requirements are intended to reduce the need for long descriptions and repetition in the measured works section of the bills of quantities. It is not normal for items in the preambles to be directly priced. The cost of complying with them is usually included within the rates set against the individual items included in the measured works section.

Measured works

The measured works section comprises a schedule of quantified descriptions of the constituent parts of a building project (including services installations and external works) compiled in accordance with SMM7R. Provided that the descriptions comply with the appropriate rules, the quantities may be approximated if they cannot be accurately determined at bill production stage. The items are normally arranged in the Uniclass work sections. The aim is to provide a comprehensive schedule of items, described and quantified according to industry-recognised conventions, which can be individually priced by contractors for the purposes of calculating their lump sum price and for fixing rates that can be used subsequently in the valuation of variations.

Provided that the Uniclass classification referencing system is used to annotate the drawings, specifications and bills of quantities, items in the bills can be directly related through that referencing system to relevant items in the specification and details on the drawings.

Formats

For tender purposes it is normal to arrange bills of quantities in the Uniclass work sections, as Example 19.1. However, it may be desirable, for various reasons, to add further information and/or to rearrange the format. Provided that the quantity surveyor appropriately annotates all items when measuring the work, the resultant bills of quantities can be readily reformatted to provide, for example:

- **Locational bills of quantities**, as Example 19.2, in which the Uniclass work sections are still adopted but the quantity for each item is broken down and allocated to a particular position within the project (e.g. a particular building, a part of a building, a house type). This format was developed to assist in the more accurate pricing of items.
- **Annotated bills of quantities**, as Example 19.3, in which the Uniclass work sections are still adopted but each item is annotated as to what it is and where it is located within the project. The annotations can be provided in a separate document, in a separate section of the bill or, most usefully, facing each item on the back of the preceding page in the bill. This format was developed as an extension of the locational bill.
- **Elemental bills of quantities**, as Example 19.4, in which the sections are based on the functional elements of the building, normally those used by the Building Cost Information Service (BCIS) (see Chapter 16), with the work in each element being arranged in Uniclass work sections. This format was developed to facilitate elemental cost analyses and as an aid to cost planning.

The early selection and appointment of the management contractor, as contemplated by the JCT MC, enables the employer to benefit from the management expertise of the contractor at an early stage in the project. When using this type of contract the project is split into discrete 'packages' that are let by the management contractor as separate 'works contracts'. The precise number of packages and their demarcation are determined by the design team and the management contractor on a project-by-project basis. The management contractor's expertise and advice is relied upon to attain the most efficient way of splitting the project into packages. Care is needed in co-ordinating the interfacing of packages to ensure that items of work are neither overlooked nor duplicated. Complete bills of quantities may be prepared for each package using any or all of the formats described above.

The examples of different formats for bills of quantities as given at the end of this chapter are not exhaustive. It is up to the design team in general and the quantity surveyor in particular to decide how best to format the bills of quantities for each project to maximise the benefit to all concerned. It is worth remembering that efficient management of documentation is a valuable aid to efficient working on site and that can be of cost benefit to the employer.

Note: Extracts from two separate sections covering three trades are set out below to allow comparison with other formats. The sections and items within them are in the same order as SMM7R.

			£
	F MASONRY **F10 Brick/Block walling**		
	Concrete blocks, BS6073, 440 mm × 215 mm, solid, keyed one side, in cement-lime mortar (1:1:6), stretcher bond		
	Walls		
A	100 mm thick : vertical	427 m²	
B	100 mm thick : vertical : curved on plan to 1500 mm radius	12 m²	
C	200 mm thick : vertical	219 m²	
	M SURFACE FINISHES **M20 Plastered/Rendered/Roughcast coatings**		
	Plaster : BS1191 part 2, undercoat 11 mm thick, finishing coat 2 mm thick : steel trowelled		
	Walls		
H	width >300 mm : 13 mm thick two coat work on brickwork or blockwork	860 m²	
	Ceilings		
J	width >300 mm : 13 mm thick two coat work on concrete	435 m²	
	Isolated columns		
K	width ≤300 mm : 13 mm thick two coat work on concrete	22 m	
	M60 Painting/Clear finishing		
	One coat sealer : two coats emulsion paint, matt finish		
	General surfaces		
T	girth >300 mm	1295 m²	
U	isolated surfaces girth ≤300 mm	22 m	

Example 19.1 Uniclass work section bills of quantities.

Note: Extracts from two separate sections covering three trades are set out below to allow comparison with other formats. The sections and items within them are in the same order as SMM7R. The quantity of each item is broken down and allocated to a particular position within the project here represented by the letters A, B & C.

			£
	F MASONRY **F10 Brick/Block walling**		
	Concrete blocks, BS6073, 440 mm × 215 mm, solid, keyed one side, in cement-lime mortar (1:1:6), stretcher bond		
	Walls		
A	100 mm thick : vertical A 340 + B 67 + C 20	427 m^2	
B	100 mm thick : vertical : curved on plan to 1500 mm radius A 0 + B 3 + C 9	12 m^2	
C	200 mm thick : vertical A 99 + B 57 + C 63	219 m^2	
	M SURFACE FINISHES **M20 Plastered/Rendered/Roughcast coatings**		
	Plaster : BS1191 part 2, undercoat 11 mm thick, finishing coat 2 mm thick : steel trowelled		
	Walls		
H	width >300 mm : 13 mm thick two coat work on brickwork or blockwork A 688 + B 129 + C 43	860 m^2	
	Ceilings		
J	width >300 mm : 13 mm thick two coat work on concrete A 350 + B 65 + C 20	435 m^2	
	Isolated columns		
K	width ≤300 mm : 13 mm thick two coat work on concrete A 15 + B 0 + C 7	22 m	
	M60 Painting/Clear finishing		
	One coat sealer : two coats emulsion paint, matt finish		
	General surfaces		
T	girth >300 mm A 1038 + B 194 + C 63	1295 m^2	
U	isolated surfaces girth ≤300 mm A 15 + B 0 + C 7	22 m	

Example 19.2 Locational bills of quantities.

Note: Extracts from two separate sections covering three trades are set out out on the facing page to allow comparison with other formats. The sections and items within them are in the same order as SMM7R. Annotations are given below.

	F MASONRY **F10 Brick/Block walling**
	Annotations
A	Non-loadbearing partitions, first floor (drawing 456/78)
B	Stores, ground and first floor and staircase enclosure walls
C	Non-loadbearing partitions, ground floor (drawing 456/77)
	M SURFACE FINISHES **M20 Plastered/Rendered/Roughcast coatings**
	Annotations
H	Loadbearing partitions, ground floor
J	Soffit of first floor
K	Columns, entrance hall
	M60 Painting/Clear finishing
	Annotations
T	Loadbearing partitions ground floor and soffit of first floor (refer to colour schedule)
U	Columns, entrance hall (refer to colour schedule)

Example 19.3 Annotated bills of quantities.

Note: Extracts from two separate sections covering three trades are set out below to allow comparison with other formats. The sections and items within them are in the same order as SMM7R. Annotations are set out on the facing page.

			£
	F MASONRY **F10 Brick/Block walling**		
	Concrete blocks, BS6073, 440 mm × 215 mm, solid, keyed one side, in cement-lime mortar (1:1:6), stretcher bond		
	Walls		
A	100 mm thick : vertical	427 m^2	
B	100 mm thick : vertical : curved on plan to 1500 mm radius	12 m^2	
C	200 mm thick : vertical	219 m^2	
	M SURFACE FINISHES **M20 Plastered/Rendered/Roughcast coatings**		
	Plaster : BS1191 part 2, undercoat 11 mm thick, finishing coat 2 mm thick : steel trowelled		
	Walls		
H	width >300 mm : 13 mm thick two coat work on brickwork or blockwork	860 m^2	
	Ceilings		
J	width >300 mm : 13 mm thick two coat work on concrete	435 m^2	
	Isolated columns		
K	width ≤300 mm : 13 mm thick two coat work on concrete	22 m	
	M60 Painting/Clear finishing		
	One coat sealer : two coats emulsion paint, matt finish		
	General surfaces		
T	girth >300 mm	1295 m^2	
U	isolated surfaces girth ≤300 mm	22 m	

Example 19.3 (*Cont.*)

Note: Extracts from two separate elements covering three trades are set out below to allow comparison with other formats. The elements are set out in the same order as is used by BCIS; sections and items within each element are in the same order as SMM7R.

			£
	2.G INTERNAL WALLS AND PARTITIONS **F10 Brick/Block walling**		
	Concrete blocks, BS6073, 440 mm × 215 mm, solid, keyed one side, in cement-lime mortar (1:1:6), stretcher bond		
	Walls		
A	100 mm thick : vertical	427 m^2	
B	100 mm thick : vertical : curved on plan to 1500 mm radius	12 m^2	
C	200 mm thick : vertical	219 m^2	
	3.A WALL FINISHES **M20 Plastered/Rendered/Roughcast coatings**		
	Plaster : BS1191 part 2, undercoat 11 mm thick, finishing coat 2 mm thick : steel trowelled		
	Walls		
H	width >300 mm : 13 mm thick two coat work on brickwork or blockwork	860 m^2	
	M60 Painting/Clear finishing		
J	One coat sealer : two coats emulsion paint, matt finish		
	General surfaces		
K	girth >300 mm	860 m^2	
L	isolated surfaces girth ≤300 mm	22 m	
	3.C CEILING FINISHES **M20 Plastered/Rendered/Roughcast coatings**		
	Plaster : BS1191 part 2, undercoat 11 mm thick, finishing coat 2 mm thick : steel trowelled		
	Ceilings		
R	width >300 mm : 13 mm thick two coat work on concrete	435 m^2	
	M60 Painting/Clear finishing		
	One coat sealer : two coats emulsion paint, matt finish		
	General surfaces		
S	girth >300 mm	435 m^2	

Example 19.4 Elemental bills of quantities.

Chapter 20
Sub-contractors

Sub-contractors

The SBC is drafted on the premise that the contractor will be wholly responsible for carrying out and completing the works. However, as sub-contracting is customary in the construction industry, the SBC makes provision for such arrangements by recognising two basic categories of sub-contractors to whom a part of the works may be sub-let; there are those sub-contractors that the contractor uses of his own volition (commonly referred to as 'domestic sub-contractors') and those sub-contractors that the contractor is directed to use by the employer (commonly referred to as 'listed sub-contractors' or 'named sub-contractors').

- **Domestic sub-contractors** Provided that the contractor obtains the written consent of the architect/contract administrator, which consent must not be unreasonably delayed or withheld, he may sub-let work to any person of his choice and such a person is referred to as a domestic sub-contractor.
- **Listed sub-contractors** The employer may require certain items of work to be executed by one of not less than three persons named in the contract documents. In such a case the contractor must sub-let that work to one of the named persons and such person is also referred to as a sub-contractor.

It is important to note that the contractor remains wholly responsible for carrying out and completing the work whether or not it is sub-let to a sub-contractor.

Specialist sub-contractors

As buildings become more complex and the systems within them more sophisticated, early decisions have to be made on various elements of them before the design team are able to proceed with the detailed design. For example, much thought must be given to the detail of a structural frame before the cladding of it can be designed; mechanical and electrical systems are frequently complex and make considerable demands on space for both plant and distribution that can affect storey heights and floor layouts. Thus it is essential for decisions relating to a number of functional elements of the building, including the selection of the principal specialist sub-contractors, to be made very early in the design process. Sub-contractors selected in this way are invariably required to contribute to or be responsible for some of the detailed design. This frequently happens in the case of structural steelwork, external wall cladding, mechanical and electrical systems, lifts and escalators.

This early involvement of sub-contractors in the design process creates a special relationship between them and the employer and in due course leads to the situation in which the employer requires the contractor to enter into a contract with the selected sub-contractor. The contractual arrangements for this sub-contract require careful consideration.

Early JCT standard forms of contract made special provisions for sub-contractors selected (nominated) in this way. Unfortunately those provisions had many short-comings, both in terms of the contractual relationship of the parties and in the responsibility for any design input, giving rise to disputes which led to arbitration and litigation. Although the JCT refined its procedures for nominating sub-contractors through standard forms and tendering procedures, employers and their advisors used to see the allocation of risk to the employer as a significant and unjustified burden. The use of the nominating procedures in JCT standard forms diminished to such an extent that they were omitted from the SBC when the JCT revised JCT 98.

Design by the sub-contractor

The SBC contemplates that part of the design will be undertaken by the contractor. Where there is a requirement for the contractor to design part of the works the SBC contains provisions for a Contractor's Designed Portion (CDP). Where the contractor is required to design the whole of the works the appropriate form of JCT contract to use is the Design and Build (DB) Contract.

Whether or not the main contract contemplates design by the main contractor, design input is frequently required from sub-contractors. Where such design input is properly planned, co-ordinated with the other production information and made available to the main contractor on time, as good practice dictates, there are seldom any practical problems. However, when this does not happen and disruption and/or delays result from a deficiency in the sub-contractor's design the employer is likely to suffer loss of time and money. The employer's remedy is to recover any such losses from the main contractor, who would undoubtedly seek redress from the sub-contractor concerned.

The SBC and sub-contract agreements

The JCT produce two Standard Building Sub-Contracts for use with each of the three versions of the SBC (SBC/Q, SBC/AQ and SBC/XQ). One version (the SBCSub) does not contain provisions for sub-contract design and the other version (the SBCSub/D) does, via a 'sub-contractor's designed portion'. The SBCSub/D design provisions are much the same as those under a contractor's designed portion in the SBC.

Each of the two sub-contracts comprises two documents:

- SBCSub comprises:
  - SBCSub/A the agreement
  - SBCSub/C the conditions
- SBCSub/D comprises:
  - SBCSub/D/A the agreement
  - SBCSub/D/C the conditions

The agreement document to each of the two contracts records the variable parts of the contract such as the articles of agreement, recitals, articles, sub-contract particulars, attestation, schedule of information and supplemental particulars. As one might glean from the list above, and in contrast to the SBC, which is a single document, the SBC sub-contracts are split into two documents, one document being the formal agreement and the other being the contract conditions. The agreement document covers the articles of agreement, recitals, attestation and specific contract variables, such as the sub-contract particulars, schedule of information and supplemental particulars. As one might expect, the contract conditions document records how the contract should be administered.

When using SBCSub/D it is integral to the agreement that the sub-contractor warrants to exercise reasonable skill and care in the:

- design of the sub-contract works
- selection of materials and goods for the sub-contract works
- satisfaction of any of the contractor's requirements.

The limitation of the sub-contractor's design liability is, broadly speaking, the same as that limiting a main contractor's liability under a contractor's designed portion of the SBC. Clause 2.13.1 states that

> *'the Sub-Contractor shall in respect of any inadequacy in such design have the like liability to the Contractor, whether under statute or otherwise, as would an architect or, as the case may be, other appropriate professional designer holding himself out as competent to take on work for such design who, acting independently under a separate contract with the Contractor, has supplied such design for or in connection with works to be carried out and completed by a building contractor who is not the supplier of the design.'*

SBC provisions under the main contract

The main contract between the main contractor and the employer, the SBC, provides for a number of mandatory conditions that all sub-contracts must contain. These, broadly speaking, comprise requirements that:

- The sub-contractor's employment under the sub-contract must terminate immediately if the main contractor's employment is terminated under the main contract.
- No unfixed materials and goods delivered to, placed on or adjacent to the works and intended for use in the works, shall be removed without first having the main contractor's consent to such removal and the main contractor must himself receive consent from the architect/contract administrator before consenting to the removal of such materials and goods.
- Where the value of any unfixed materials and goods have been included in an interim certificate for payment to the main contractor and paid to him, those materials and goods shall become the property of the employer and the sub-contractor shall not dispute this.
- Where the main contractor pays the sub-contractor for any unfixed materials and goods before their value is included in an interim certificate, upon payment those materials and goods shall become property of the main contractor.
- The sub-contractor shall grant the rights of access to workshops or other premises where work is being prepared for the contract to the architect/contract administrator or any person authorised by him.
- If the main contractor fails to pay the amount properly due to the sub-contractor by the final date for payment, the contractor shall, in addition to the amount properly due, pay simple interest thereupon.
- Where applicable, the sub-contractor must deliver a warranty in favour of the employer within 14 days of receiving a written request from the contractor. This is intended to give the employer/funder/purchaser/tenant a right of recourse against the sub-contractor (for the sub-contractor's obligations) should the main contactor cease to exist.

Chapter 21
Obtaining Tenders

Introduction

It is frequently desirable to appoint the contractor at an early stage in the design process but to do so necessitates various modifications to traditional tendering procedures (see Chapters 3 and 15). Having said that traditional tendering procedures are still widely adopted and are therefore considered in more detail in this chapter.

There have so far been three significant published codes relating to the selection of contractors. In January 1996 the National Joint Consultative Committee for Building (NJCC) published *Code of Procedure for Single Stage Selective Tendering* in collaboration with the Scottish Joint Consultative Committee and The Joint Consultative Committee for Building, Northern Ireland. The NJCC was voluntarily dissolved on 27 June 1996 but there is still a tendency within the construction industry to use its code. In May 1997 the Construction Industry Board (CIB) published *Code of Practice for the Selection of Main Contractors*. The CIB ceased to exist on 29 June 2001 but its code, although it does not seem to be as widely adopted as the NJCC code, continues to be published by Thomas Telford Publishing. In July 2002 the JCT published *Practice Note 6 Main Contract Tendering* to provide model forms of tender which, at the time of writing, is due to be revised and republished in 2006 to complement the recently published 2005 suite of contracts. Whilst the practice note is a fairly recent publication, it does not require any new overly excessive procedures to those of the NJCC code upon which the practice note is based.

The NJCC, CIB and JCT publications all codify the principles of good practice relating to the selection of contractors and although the NJCC and JCT publications differ in detail from the CIB publication they all cover the following topics:

- tender list
- tendering procedure
    - preliminary enquiry
    - tender documents and invitation
    - tender period

The primary aim ... is to produce a list of contractors all of whom are capable of completing the job satisfactorily

- — tender compliance
- — late tenders
- tender assessment
  - — opening tenders
  - — examination and adjustment of the priced document
  - — negotiated reduction of tender
- notification of results.

The primary aim of selection is to produce a list of contractors all of whom are capable of completing the project satisfactorily. The final choice of contractor can then be the one submitting the lowest or most economically advantageous bona fide tender.

Tender list

The cost to a contractor of preparing a tender is high and is therefore reflected in the cost of building. If a large number of contractors were afforded the opportunity to tender for a project, the total amount of abortive costs would increase proportionately by the number of unsuccessful contractors submitting tenders. Whilst the total abortive costs might not impact directly on the employer, the abortive costs of unsuccessful tenders are spread across projects to which successful tenders have been submitted. It therefore follows that open tendering increases the costs of projects unnecessarily and should, if at all possible, be avoided.

In order to minimise abortive tendering costs the design team should ensure, as early in the design process as possible, that there exists a list of suitable contractors from which those invited to tender will be selected during stage H of the plan of work (see Example 16.1). Those employers who build frequently will normally maintain a list of suitable contractors. Such a list should be reviewed on a regular basis so that contractors who have not performed well or who have otherwise become unsuitable can be removed and those not on the list who meet the relevant criteria can be added to it. When there is no list of approved contractors one should be compiled from firms known to the employer and the design team and/or from those who respond to press advertisements. Such advertisements need to be carefully worded and must at least indicate the size, nature and location of the project and the date when the tender documents will be ready.

It is worth noting here that European Union procurement rules set down contractor selection and contract award procedures applicable to public authorities in respect of contracts above certain specified values. Such contracts must be the subject of notices in the *Official Journal of the European Communities*. Provided that the contracting authority opts for what is termed the 'restricted procedure' then, by use of the criteria stated in the notice, it may select those who will be invited to tender from the contractors who respond. The final list of contractors to be invited to tender should comprise no more than six plus two reserves. The main criteria for their selection should include:

- adequacy of available resources
- adequacy of technical and management structure
- financial stability and insurance cover
- health and safety record
- quality of work and adequacy of quality control
- performance record.

Preliminary enquiry

About a month before the tender documents are due to be despatched the selected contractors should be sent a letter containing as much of the information listed below as possible and asking whether they wish to tender for the project:

- project name, function and general description
- employer
- design team
- location of site
- approximate cost range
- number of tenderers
- form of contract and whether to be executed under hand or as a deed
- any sub-contractors for major items
- anticipated date of possession

- contract period
- anticipated date for despatch of tender documents
- tender period
- period for which tender is to remain open.

After the latest date for response to the preliminary enquiry the list of tenderers should be finalised and those included on the list notified. At the same time any contractors who asked to tender by responding to an advertisement and who are not included on the tender list should be informed of this.

Tender documents and invitation

On the date named in the preliminary enquiry the tender documents should be sent to or made available for collection by the selected tenderers. The tender documents should comprise:

- a checklist of all tender documents
- instructions to tenderers
- two copies of the drawings, schedules and specification
- two copies of the bills of quantities or pricing schedules
- two copies of the health and safety plan
- two copies of the form of tender
- suitably addressed envelopes.

One envelope should be for the return of the tender, endorsed with the word 'Tender' and the name of the project together with the latest date and time for the tender return. Another envelope should be for the return of the priced bills or schedules in support of the tender (if they are to be returned with the tender), endorsed with the project name and the words 'Priced Bills' or 'Priced Schedules' together with a space for the tenderer's name.

The instructions to tenderers referred to above should include:

- where and by when to submit the tender
- any information required to be submitted with the tender, such as method statement, quality control resources and programme of work
- method of packaging and identifying the tender
- method for dealing with any queries relating to the tender documents
- method of dealing with any errors or inconsistencies in the tender documents discovered after they have been issued
- whether or not alternative proposals are acceptable if accompanying a compliant tender

- notes relating to any contractor's designed portion
- required period of validity for the tender
- anticipated tender acceptance date
- arrangements for inspecting additional information
- arrangements for visiting the site
- whether the employer will accept the lowest or any tender
- tender assessment criteria
- method of handling errors in tenders
- method of communication of tender results.

Contractors should of course acknowledge receipt of the documents and confirm that they are still able to provide a bona fide tender by the due date or, if for some reason they are unable to do so, they should return the documents immediately.

Tender period

The time required by a contractor to prepare a tender is dependent on both the size and complexity of the project. Whilst it is generally accepted that the minimum tender period should be four weeks, a longer period will be required in some instances. If realistic prices are to be tendered it is imperative that tenderers be given sufficient time in which to obtain competitive prices from their suppliers and sub-contractors, to comply with the requirements of CDM94 and to formulate their bids properly.

Tender compliance

In order to achieve fair competitive tendering it is essential that any unauthorised amendments to or qualifications of the tender documents by a tenderer render the tender non-compliant and subject to rejection, although the tenderer should be given the opportunity to withdraw the amendments/qualifications and stand by his tender. It is also essential that unsolicited alternative bids, either in terms of price or time, are considered non-compliant and rejected.

Late tenders

The latest time and date for the submission of tenders is best established when the tender documents are first made available. This date may, for a variety of reasons, have to be extended, in which case all tenderers must be notified. Any tender received after the latest time and date fixed for their receipt should not be admitted to the competition and is best returned unopened to the tenderer.

Opening tenders

Tenders should be opened as soon as possible after the published time for their receipt. It is as well if at least two people are present. Upon opening, each tender should be countersigned by two of those present and the tenderer and tendered amount should be entered on a list which, when complete, should also be signed by two of those present.

Examination and adjustment of the priced document

The lowest tenderer should be requested to provide a priced document (bills of quantities or pricing schedule) in support of the tender if this was not required to be submitted with it. The quantity surveyor should examine the priced document in support of the lowest tender to determine that any amendments notified during the tender period have been properly included and to detect any errors in computation. Any errors found are dealt with in the manner prescribed in the instructions to tenderers; it is important that the manner is laid down from the outset rather than left until the return of tenders.

The most frequently adopted methods for dealing with errors in computation are:

- The tenderer is advised of the errors and is given the opportunity to confirm or withdraw his offer. If he withdraws, the examination process is repeated with the next lowest tenderer. If he is prepared to stand by his tender, the pricing document is endorsed to the effect that all rates and prices, excluding those relating to provisional sums and prime cost sums, be considered increased or decreased in the same proportion as the corrected total of the priced items exceeds or falls short of the uncorrected total of such items. It is these adjusted rates and prices that will subsequently be used for valuing variations.

- The tenderer is advised of the errors and is given the opportunity to amend his tender to correct genuine errors or to withdraw his offer. If he amends and the amended tender is no longer the lowest, or if he withdraws, the examination process is repeated with the lowest tenderer.

When a tender appears to be free from error, or when in spite of error the contractor is prepared to stand by his tender, or when a tender is still lowest after amendment, such a tender should be recommended for acceptance. However, there are occasions when the tender instructions indicate that price will not be the only criterion by which the offers are to be assessed and that such factors as the tenderer's proposed solutions to specific problems, his proposed method of working or his proposed programme will be equal to or of greater importance than price. When this is the case the design team evaluate the bids against the relevant criteria and may conclude that

... is prepared to stand by his tender

the lowest price offered is not necessarily the most appropriate one to recommend for acceptance.

Negotiated reduction of a tender

Good practice dictates that a contractor's tendered price should not be changed on an unamended scope of works except for the correction of genuine errors, as noted above. However, should the lowest tender exceed the employer's budget, it is permissible for a reduced price to be negotiated with the lowest tenderer on the basis of agreed changes in the specification and/or quantity of the work. If such negotiations fail then, and only then, is it permissible to commence negotiations with the next lowest tenderer.

Notification of results

Immediately after the tenders have been opened all but the lowest three tenderers should be notified that their tenders have not been successful. At the same time the second and third lowest should be advised that they are not the lowest but that they may be approached again should the lowest tenderer withdraw his offer. As soon as a decision has been made to accept a tender, all unsuccessful tenderers should be so notified and any submitted priced documents should be returned to them unopened.

The employer should be encouraged to decide as quickly as possible which tender to accept, for it is of the utmost importance to contractors to know whether or not their tenders have been successful. Once the contract has been let it is good practice to provide each tenderer with a complete list of the tender prices received. Whilst this does not disclose who tendered which amount, it does enable each tenderer to see how they stood in relation to the rest.

Tender analysis

Frequently it will be required for the quantity surveyor to analyse the successful tender and in any event it is good practice. Traditionally the analysis is based on the functional elements of the building (see Chapter 16) to enable the quantity surveyor to revise the cost plan to reflect the tendered prices in order to establish a proper basis for post-contract cost control (see Chapter 27) and for the purposes of adding to the industry's historical cost data bank.

More and more projects are dependent upon third-party funding and government-sponsored financial incentives for their economic viability. The quantity surveyor is therefore frequently required to provide financial information, normally gleaned from tender analysis, to satisfy funders, to support grant applications and to enable the employer to take advantage of capital allowances and other forms of tax relief offered by HM Revenue and Customs (see Chapter 31).

Part IV

Contract Administration

Chapter 22
Placing the Contract

The placing of the contract is a relatively simple, routine matter but the events that immediately precede it and those that immediately follow it are far from simple and are certainly of great importance.

Preparing and signing the contract documents

If the date of possession of the site by the contractor and the date for completion of the works have not previously been decided they certainly should be agreed with the contractor whilst the contract documents are being prepared for signing. At the same time it is as well to make arrangements for the initial site meeting and to ensure that the contractor has no valid objection to any of the proposed sub-contractors and suppliers (see Chapter 20).

The contract documents, which should be so labelled, will normally comprise the articles of agreement, the conditions of contract, the drawings showing the work to be done, the priced bills of quantities or the priced document and any post-tender negotiation documentation. In preparing the SBC for signing it is necessary to complete the articles of agreement and the contract particulars at the beginning together with schedule six relating to the form of bonds at the end. It is also necessary to make the following deletions and amendments in the text of the conditions of contract:

- Clause 1.1 – amend as necessary the definition of public holidays if different public holidays are applicable.
- Clause 1.12 – amend as necessary when the parties to the contract do not wish the law applicable to the contract to be English law.
- Clause 6.7 – always delete two of these insurance options. Option A is applicable to the erection of new buildings when the contractor is required to take out the insurance. Option B is applicable to the erection of new buildings when the

employer is required to take out the insurance. Option C is applicable to alterations and extensions to existing structures in which case the contract provides only that the employer is required to take out the insurance. Footnotes to these clauses remind users that it is sometimes not possible to obtain insurance in the precise terms required by SBC, in which case the conditions must be amended accordingly.

The information necessary for the completion of the articles of agreement and the contract particulars together with details of the above noted deletions and amendments should have been included in the tender documents. It is imperative that no other insertions, deletions or amendments are made to the SBC without the prior agreement of the contracting parties.

The SBC's predecessor, JCT 98, had three supplements which, if relevant, had to be included in the documentation. These were:

- Sectional Completion Supplement
- Contractor's Designed Portion Supplement
- Composite Contractor's Designed Portion and Sectional Completion Supplement.

The SBC now incorporates sectional completion and contractor's designed portion as standard text. It is worth noting here that sectional completion is not to be confused with partial possession by the employer. Sectional completion relates to a specific requirement on the part of the employer, recorded in the contract documents, for the works to be completed in discrete, phased sections by pre-determined dates. In contrast, partial possession by the employer is prescribed by SBC clauses 2.33 to 2.37 which provide that the employer may, with the consent of the contractor, take possession of a part or parts of the works or section at any time prior to practical completion. The same clauses also cover the practical considerations such as practical completion of the part(s) taken over, defects and insurance.

Sectional completion

Sectional completion under the SBC is for use when the works are to be completed in phased sections, of which the employer will take possession upon practical completion of each. If sectional completion is to be used, it is necessary for the works to be divided into the sections in the tender documents at tender stage. The number of sections that a project may be divided into is, in theory, infinite. However, in practical terms, the project will be influenced by physical and commercial considerations. Each section is treated as having its own date of possession and dates for completion of section. In effect, each section is treated as being a mini project in its own right. Should a contractor fail to complete a specific section by that section's date for completion, the contractor will be liable for the amount of liquidated damages stated in the contract particulars. The contract particulars must state the amount of damages applicable to each section.

Contractor's designed portion

The contractor's designed portion (CDP) is for use when the contractor is required to complete the design of a portion of the works. The use of the contractor's designed portion option creates the following additional contract documents:

- **The Employer's Requirements** This document shows and describes the requirements of the employer in respect of that portion of the works to be designed by the contractor.
- **The Contractor's Proposals** This document shows and describes the contractor's proposals for the design of the relevant portion of the works to meet the employer's requirements.
- **The CDP Analysis** This document is an analysis of the portion of the contract sum relating to the portion of the works to be designed by the contractor.

Contractor's Designed Portion ...

It will be noted that these documents are the same as those in use with the JCT's design and build contract. Whist the references to and obligations surrounding the CDP are spread throughout the contract, the principal clauses comprise the following:

Recitals

Seventh	the Works include the design and construction of . . .
Eighth	the Employer has supplied the Contractor documents . . . for the design and construction of the Contractor's designed portion
Ninth	in response to the Employer's Requirements the Contractor has supplied the Employer: . . .
Tenth	the Employer has examined the Contractor's Proposals and, subject to the Conditions, is satisfied that they appear to meet the Employer's Requirements

Articles

1 Contractor's obligations

Clause

2.2	Contractor's Designed Portion
2.3	Materials, goods and workmanship
2.9.2 , 2.9.3 & Schedule 1	Construction information and Contractor's master programme
2.14.4	Contract Bills and CDP Documents – errors and inadequacy
2.15.5	Notification of discrepancies etc.
2.16	Discrepancies in CDP Documents
2.17	Discrepancies from Statutory Requirements
2.19	Design liabilities and limitation
2.20	Errors and failures – other consequences
2.40	Contractor's Design Documents – as-built drawings
2.41	Copyright and use

3.7.2	Consent to sub-letting
3.10.3	Compliance with instructions
3.14.3	Instructions requiring Variations
5.8	Contractor's Designed Portion – Valuation
8.7.2.2	Consequences of termination under clauses 8.4 to 8.6

The contractor's main obligation under a contractor's designed portion is to complete the design portion in accordance with the contract drawings and contract bills. Depending on the level of detail contained in the employer's requirements, the contractor must develop the specification in respect of the selection of materials and goods and the standard of workmanship.

Executing the contract

The tender documents should record, at tender stage, whether the agreement is to be executed under hand or as a deed. The Limitation Act 1980 provides that actions for breach of contract can be commenced within six years of the breach in respect of a contract executed under hand and within 12 years of the breach in respect of a contract executed as a deed.

The attestation clauses in the SBC make provision for the agreement to be executed under hand or as a deed both by a company or other body corporate and by an individual. Irrespective of how the agreement is executed, both parties to the contract should initial each and every contract document. They should both also initial all amendments made to the SBC when used.

Performance bonds and parent company guarantees

A performance bond is a three-party agreement between the contractor, a surety and the employer that provides for the surety to pay a sum of money to the employer in the event of default by the contractor. The value of the bond normally provided is equal to 10% of the original contract sum. As the raising of a bond invariably imposes a financial burden on the contractor in terms of borrowing, it is considered to be good practice for the bond to be released coincidentally with the achievement of a particular event, e.g. practical completion or the issue of the certificate making good.

The contractor will normally obtain such a bond from an insurance company or a bank. However, where the contractor is a subsidiary of a large organisation it may well be that a guarantee from the parent company will suffice instead of a bond. The bond holder and the terms of the bond or guarantee must be approved by the employer. A typical performance bond (Example 22.1) and a typical parent company guarantee (Example 22.2) are reproduced at the end of this chapter. Whether or not a bond or guarantee is required must be determined prior to tender stage and the appropriate entry made in the tender documents so that the tenderers can make due allowance for all associated costs in their tendered prices.

Collateral warranties

Collateral warranties in general and those to be provided by consultants in particular are considered in Chapter 2, where mention is made of the standard warranty agreements published by BPF and CIC.

Like consultants, main contractors are also frequently required to enter into collateral warranty agreements and are always required to do so when those entered into by the consultants are the BPF agreements CoWa/F and CoWa/P&T or the CIC agreements CIC/ConsWa/F and CIC/ConsWa/P&T, under which the responsibilities of the consultants are on the basis that the main contractor will provide contractual undertakings to the funding institution, purchaser or tenants to similar effect as those provided by the consultants. Section seven of the SBC contains provision for JCT warranties to be entered into and refers to part two of the contract particulars. JCT warranties comprise:

- CWa/F – Contractor Collateral Warranty for a Funder
- CWa/P&T – Contractor Collateral Warranty for a Purchaser or Tenant.

Third party rights

The Contracts (Rights of Third Parties) Act 1999 referred to in Chapter 2 affords parties to a contract a viable alternative to collateral warranties. The SBC expressly affords parties to the contract the choice of either third party rights under the Act or collateral warranties, should the need to circumvent the doctrine of privity of contract be required. The SBC envisages that those requiring such rights are those who would ordinarily require collateral warranties – funders, purchasers or tenants. Section seven of the SBC covering third party rights requires that, in respect of purchasers or tenants, the class or description of those with the benefit of the rights is made known and the contract particulars state for which part of the works such rights are required.

Issue of documents

The SBC requires that the contract documents remain in the custody of the employer so as to be available at reasonable times for inspection by the contractor. It also requires that immediately after the contract has been executed the architect/contract administrator must provide the contractor with:

- one copy of the contract documents certified on behalf of the employer
- two further copies of the contract drawings
- two copies of the unpriced bills of quantities or specification/schedules of work.

The SBC also requires that as soon as possible after the execution of the contract:

- The architect/contract administrator shall provide the contractor with two copies of any descriptive schedules or other similar documents necessary for use in carrying out the works.
- The contractor shall provide the architect/contract administrator with two copies of his master programme for the execution of the works.
- The contractor shall provide the architect/contract administrator with two copies of the contractor's design documents and related calculations as are reasonably necessary to explain or amplify the contractor's proposals
- The contractor shall provide the architect/contract administrator with two copies of all the levels setting out dimensions which the contractor has prepared for completing any CDP.

The contractor's design documents are required to be provided as and when necessary in accordance with the contractor's design submission procedure and clause 2.9.3 and Schedule One of the SBC.

Insofar as they are relevant to the project the following documents will also have to be provided to the contractor by the design team:

- party wall agreements
- condition surveys of adjoining properties
- conditional planning approvals
- tree preservation orders
- building control approval and notices to be served during the course of the works
- procedures and estimates for works to be carried out by statutory and local authorities.

The contractor is required by the SBC to keep one copy of each of the following documents on site so as to be available to the architect:

- contract drawings
- the unpriced bills of quantities or specification/schedules of work
- CDP documents
- descriptive schedules or other similar documents

- the master programme
- any further drawings or details issued to explain and amplify the contract drawings.

Insurances

The insurance provisions in the SBC are to be found in Section 6 and Schedule 3.

Clause 6.4.1 requires the contractor to take out and maintain insurance against claims for personal injury to or death of any person or for loss, injury or damage to real or personal property that arises out of or in the course of or is caused by the carrying out of the works except for claims caused by a default of the employer or those for whom the employer is responsible. The insurance in respect of the contractor's liability to third parties must also extend to meet any like claims made against the employer by third parties. Insurance in respect of injury or death to an employee of the contractor must comply with the current legislation (at the time of writing this is the Employers' Liability (Compulsory Insurance) Regulations 1989). Insurance in respect of all other claims has to be for no less cover than is stated in the contract particulars.

Clause 6.5.1 also provides the employer with the option to require the contractor, by way of an architect's instruction, to insure (in the names of the employer and the contractor) against any expense, liability, loss, claim or proceedings incurred by the employer as a result of injury or damage to any property, other than the works and any site materials for the works, caused by:

- collapse
- subsidence
- heave
- vibration
- weakening or removal of support
- lowering of ground water

arising out of or in the course of or by reason of the carrying out of the works. This would exclude injury or damage:

- caused by the default of the contractor
- caused by errors or omissions in designing the works
- which can reasonably be foreseen to be inevitable with regard to the nature of the work
- which it is the responsibility of the employer to insure under the SBC Option C (if applicable) – see below
- arising from the consequence of war and like acts
- caused by the excepted risks (as defined in SBC clause 6.8)

- caused by pollution or contamination
- resulting in damages payable by the employer for breach of contract.

The extent of cover required has to be stated in the contract particulars, but as this particular insurance is instigated by an architect's instruction the cost to the contractor of taking it out and maintaining it is added to the contract sum as an extra.

Clause 6.7 deals with insurance of the works and Schedule Three provides three alternatives:

- **Option A** Where the works comprise new buildings and the parties agree that the contractor is required to take out and maintain a joint names policy for all risks insurance of the works.
- **Option B** Where the works comprise new buildings and the parties agree that the employer is required to take out and maintain a joint names policy for all risks insurance of the works.
- **Option C** Where the works are in or are extensions to existing structures. Here the employer is required to take out and maintain a joint names policy for loss or damage caused to the existing structures and contents by one or more of the specified perils and a joint names policy for all risks insurance of the works.

The term 'joint names policy' is expressly defined in the SBC at clause 6.8 as being 'a policy which includes the Employer and the Contractor as composite insured and under which the insurers have no right of recourse against any person named as an insured, or, pursuant to clause 6.9, recognised as an insured thereunder.' Clause 6.9 further provides that the joint names policy must either provide for recognition of each sub-contractor as an insured or include a waiver by the insurers of any right of subrogation which they may have against any such sub-contractor. Subrogation is the doctrine under which an insurer, who has paid the insured for loss or damage suffered, is entitled to sue, in the insured's name, whoever caused the loss or damage.

The amount of cover to be provided by the joint names policy is the full reinstatement value of the works plus the percentage stated in the contract particulars to cover professional fees and, in the case of works in or extensions to existing structures, plus the full value of the existing structures and of their contents owned by or for which the employer is responsible.

Clause 6.7 of the SBC under Option A.3 recognises that it is common practice for contractors to maintain annually renewable insurance policies which provide appropriate Option A all risks insurance and sets out 'deemed-to-satisfy' provisions in respect of such annual policies.

The SBC, amongst other things, defines terrorism and terrorism cover and provides the employer with the option of determining the employment of the contractor or of requiring the contractor to complete the works at the employer's expense if insurers withdraw terrorism cover during the currency of the contract. These provisions were previously contained in JCT Amendment TC/94, issued April 1994. Following terrorist attacks on the UK mainland, particularly the IRA bombings in the City of London, which caused billions of pounds worth of damage, insurance against terrorism had become unavailable. In order to comply with the SBC requirement to maintain cover, the government agreed to act as 'reinsurer of last resort'.

Construction insurance is an extremely complex subject which is not helped by the fact that available policies do not incorporate standard wording. Consequently the employer should always be advised to consult a specialist insurance adviser. Once insurances are in place it is important that they are maintained and renewed when necessary. It is therefore recommended that the design team checks that insurances are maintained and renewal premiums are paid when due. It is advisable to keep copies of premium receipts or brokers' confirmatory letters on file.

THIS BOND is made the .... day of .... 20 ......
BETWEEN ....
of ....
(hereinafter called 'the Contractor') of the first part.
and ....
of ....
(hereinafter called 'the Surety') of the second part.
and ....
of ....
(hereinafter called 'the Employer') of the third part.

1. By a Contract dated .... made between the Contractor and the Employer (hereinafter called 'the Contract') the Contractor has agreed to carry out the Works specified in the Contract for the sum of £ ....

2. The Contractor and the Surety are hereby jointly and severally bound to the Employer in the sum of £ .... (not exceeding ten per cent of the original Contract Sum) which sum shall be reduced by an amount equal to ten per cent of the value of any part or parts of the Works taken into possession of the Employer under the provisions of the Contract *(or of any Section of the Works upon the Architect/Contract Administrator certifying practical completion of that Section) provided that if the Contractor shall, subject to Clause 5 hereof, duly perform and observe all the terms, conditions, stipulations and provisions contained or referred to in the Contract which are to be performed or observed by the Contractor or if on default by the Contractor the Surety shall, subject to Clause 3 hereof, satisfy and discharge the damage sustained by the Employer thereby up to the amount of this Bond then this agreement shall be of no effect.

* *The wording in brackets should be deleted unless Sectional Completion applies.*

3. If the Contractor has failed to carry out the obligations referred to in Clause 2 hereof then written notice requiring the Contractor to remedy his failure, where possible, shall be given, and if the Contractor fails so to do or repeats his default or if the Contractor's employment under the Contract is determined in accordance with clause 8.4 of that Contract the Employer shall be entitled to call upon the Surety in accordance with Clause 2.

4. Any alteration to the terms of the Main Contract or any variations required under the Contract shall not in any way release the Surety from its obligations under this agreement.

5. The Contractor and the Surety shall be released from their respective liabilities under this agreement upon the date of the Practical Completion of the Works as certified by the Architect/Contract Administrator appointed under the Contract.

IN WITNESS whereof the parties hereto have executed this Document as a Deed the day and year first before written.

Contractor:	....
Signed by (insert name of Director) and	Director
(insert name of Company Secretary or	....
second Director) for and on behalf of	Director/Company Secretary
....	
Surety:	....
Signed by (insert name of Director) and	Director
(insert name of Company Secretary or	....
second Director) for and on behalf of	Director/Company Secretary
....	
Employer:	....
Signed by (insert name of Director) and	Director
(insert name of Company Secretary or	....
second Director) for and on behalf of	Director/Company Secretary
....	

Example 22.1 Performance bond.

THIS AGREEMENT is made the .... day of .... 20 ......
BETWEEN ....
whose registered office is at ....
(hereinafter called 'the Guarantor') of the one part and ....
whose registered office is at ....
(hereinafter called 'the Employer') of the other part,
WHEREAS

A. ....
(hereinafter called ....
a subsidiary of the Guarantor has entered into a Contract with the Employer of even date to carry out certain works at ....
(hereinafter called 'the Contract')

B. The Guarantor has agreed to guarantee the due performance by ....
.... of the Contract in the manner hereinafter appearing:

NOW THE GUARANTOR HEREBY AGREES WITH THE EMPLOYER as follows:

1. If .... (unless relieved from the performance by any clause of the Contract or by statute or by the decision of a tribunal of competent jurisdiction) shall in any respect commit any breach of its obligations or duties under the Contract then the Guarantor will procure the remedying of any breach of .... obligations failing which the Guarantor will pay to the Employer and/or his assigns the amount (subject as hereinafter provided) of all losses, damages, costs and expenses which may be incurred by the Employer by reason of any default on the part of .... in performing its obligations and duties or otherwise to observe and comply with the provisions contained in the Contract to the extent that such losses, damages, costs and expenses are or would otherwise be recoverable by the Employer from .... under the said Contract.

2. The Guarantor further agrees with the Employer that it shall not in any way be released from liability hereunder by any alteration in the terms of the Contract made by agreement between the Employer and .... or in the extent or nature of the Works to be constructed, completed and maintained thereunder or by any allowance of time or forbearance or forgiveness in or in respect of any matter or thing concerning the Contractor or by any other matter or thing whereby (in the absence of this present provision) the Guarantor would or might be released from liability hereunder, and for the purposes of this Agreement any such alteration shall be deemed to have been made with the consent of the Guarantor and Clause 1 hereof shall apply in respect of the Contract so altered. Provided always that the amount of the Guarantor's liability under this Agreement shall be no greater than the amount which would have been recoverable against the Contractor by the Employer under the Contract and the same limitation periods fixed by statute which apply to the Contract shall equally apply to this Agreement.

IN WITNESS whereof the parties hereto have executed this Document as a Deed the day and year first before written.

Guarantor:	....
Signed by (insert name of Director) and	Director
(insert name of Company Secretary or	....
second Director) for and on behalf of	Director/Company Secretary
....	
Employer:	....
Signed by (insert name of Director) and	Director
(insert name of Company Secretary or	....
second Director) for and on behalf of	Director/Company Secretary
....	

Example 22.2 Parent company guarantee.

Chapter 23
Rights, Duties and Liabilities under the SBC

Introduction

Standard forms of construction contract are far more than a simple agreement to undertake and complete a construction project. They prescribe and allocate important rights, duties and liabilities to different members of the building team for most, if not all, situations commonly encountered during the process.

With the exception of domestic construction projects – domestic alterations and extensions – most projects are incredibly complex undertakings and require significant contribution from the many professions making up the building team. It is the harnessing of the building team and, in particular, their skills which ultimately contributes towards the success or the failure of the project. Nevertheless, whilst standard contracts allocate rights, duties and responsibilities to building team members, not all team members have express powers. Those without powers must communicate via those administering the contract. Failure to perform the allocated responsibilities can dramatically increase the chance of a dispute arising.

Set out below in the rest of this chapter is a schedule formulating and allocating important rights, duties and liabilities based on the SBC/Q.

The employer

Recitals

First	Duty to state the nature of intended works.
Third	Duty to sign the contract drawings and the contract bills.
Fourth	Duty to give status for the purposes of the Construction Industry Scheme.
Fifth	Optional duty to provide an information release schedule.

Sixth	Duty to identify the sections of the works, if sectional completion is required, in the contract bills and/or the contract drawings or such documents identified in the contract particulars.
Seventh	Duty to state the nature of any intended contractor's designed portion (CDP).
Eighth	Duty to provide the contractor with documents showing and describing the requirements for any CDP.
Ninth	Right to receive the contractor's proposals for the design and construction of any CDP and an analysis for the portion of the contract sum to which the CDP relates.
Tenth	Duty to examine the 'contractor's proposals' and satisfy himself that they meet the 'employer's requirements'. Duty to sign or initial the employer's requirements, contractor's proposals and the CDP analysis.

Articles

2	Duty to pay the contract sum.
3	Duty to nominate and appoint a replacement architect/contract administrator, if needed.
4	Duty to nominate and appoint a replacement quantity surveyor.
5	Duty to appoint a replacement planning supervisor.
6	Duty to appoint a replacement principal contractor.
7	Right to refer a dispute to adjudication.
8	Right to refer a dispute to arbitration if the parties have agreed that it is the method for finally determining disputes.

Clauses

1.9	Right to be issued with all architect/contract administrator's certificates.
2.4	Duty to give the contractor possession of the site or, in the case of section, the relevant part of the site.
2.5	Right to defer possession of the site.
2.6.1	Right to use or occupy the site or the works, in whole or in part, with consent of the contractor.
2.6.2	Duty to pay the contractor for increased insurance premiums, if the contractor insures the works under Option A and the increase in premiums arises as the consequence of the employer's occupation under 2.6.1.

2.8.1 Duty to retain custody of the contract documents and make them available to the contractor for inspection at all reasonable times.

2.8.4 Duty not to divulge any rates or prices in the contract bills, except for the purposes of the contract.

2.10 Right to require by architect/contract administrator's instruction that setting out errors shall not be corrected.

2.11 Right to agree a variation to the information release schedule. Right and duty in respect of the agreement, which shall not be unreasonably delayed or withheld.

2.13.2 Liability in respect of the contents of the employer's requirements.

2.21 Right to be indemnified against any fees or charges legally demandable under the statutory requirements. Items that the employer might be indemnified against comprise any fees or charges incurred via statute, statutory instrument, regulation, European directive, local authority or statutory undertaker.

2.22 Right to indemnification against all claims and proceedings incurred by the contractor infringing or being held to have infringed any patent rights.

2.24 Right to ownership of unfixed materials and goods on-site.

2.25 Right to ownership of unfixed materials and goods off-site.

2.29.3 Liability in the event of deferment of giving possession of site.

2.29.5 Liability in the event that the contractor suspends performance of his obligations for non-payment.

2.29.6 Liability in respect of any impediment, prevention or default by act or omission, except to the extent that it was not caused or contributed to by the contractor.

2.32 Rights and duties in respect of liquidated damages.

2.33 Right to take partial possession of a relevant part of the works if consent is given by the contractor. Right not to have consent unreasonably withheld.

2.36 Duty to insure the relevant part on taking partial possession of the works or section.

2.38 Right to have all shrinkages and defects appearing in the rectification period made good within a reasonable time and at no extra cost.

2.39 Right to require the architect/contract administrator to instruct that defects, shrinkages and other faults shall not be made good and have an appropriate deduction from the contract sum.

2.40 Duty to be provided, without charge, as-built drawings and contractor's design information in respect of any CDP.

2.41.2 Right, which is conditional on all monies due under the contract having been paid, to have an irrevocable, royalty free, non exclusive licence to copy the contractor's design documents for any purpose relating to the works, except for reproducing the designs contained therein.

3.3	Right to appoint an individual to act as a representative. Duty to inform the contractor if and when a representative is appointed.
3.4	Right to appoint a clerk of the works.
3.5.1	Duty to appoint a replacement architect/contract administrator or quantity surveyor if and when an original appointee ceases to hold office.
3.7.2	Right to be consulted in respect of a contractor sub-letting design obligations under any CDP. Right to consent to any such sub-letting. Duty not to unreasonably withhold or delay such consent.
3.8.2	Right, with consent of the contractor, to add additional persons to a list contained in the contract bills for work described and contained therein.
3.8.3	Duty, together with the contractor, to add names to the list in 3.8.2 when the list falls below three persons.
3.9.2.1	Right to ownership of sub-contract unfixed materials and goods on site.
3.11	Right to employ others if the contractor does not comply with any architect/contract administrator's instruction.
3.18	Right to agree to allow non-compliant materials or goods to remain and to obtain an appropriate deduction from the contract sum.
3.22	Right to receive all fossils, antiquities and other objects of interest or value found during the progress of the works.
3.25	Rights, duties and liabilities in respect of complying with the CDM Regulations.
3.26	Duty to notify the contractor of the appointment of a replacement planning supervisor or a replacement principal contractor.
4.3.1.1	Right to have the contract sum adjusted by any amount agreed with the contractor in respect of variations.
4.3.2	Right to have the amounts due under certain prescribed clauses deducted from the contract sum.
4.3.3	Liability in respect of the contractor's costs incurred under the prescribed clauses.
4.7	Duties under the Construction Industry Scheme in respect of the Corporation Taxes Act 1988 if the employer is a 'contractor'.
4.8	Rights, duties and liabilities in respect of advance payments due to the contractor.
4.13	Rights, duties and liabilities in respect of payments due to the contractor.
4.14	Liability in respect of non-payment of amounts due to the contractor.
4.15	Rights, duties and liabilities in respect of final payments due to the contractor or from the contractor as the case may be.
4.17	Rights and duties in respect of payment for off-site materials or goods.
4.18	Rights and duties in respect of retention.

4.19	Rights in respect of the provision of a bond by the contractor in lieu of retention.
4.20	Right to deduct and retain the retention from interim payments.
4.23	Liability in the event that the contractor incurs direct loss and/or expense.
4.24.5	Liability in respect of any impediment, prevention or default whether by act or omission, except insofar as it was not caused or contributed to by the contractor.
5.1.2	Right to impose or vary any obligations or restrictions in regard to access to or use of the site, and in regard to working space, working hours and the order of work.
5.2.1	Rights and duties with regard to valuing variations.
5.3	Rights and duties in respect of the acceptance of a Schedule 2 quotation.
6.1 & 6.2	Rights, duties and liabilities in respect of injury or death and loss of or damage or injury to property and related insurance.
6.5	Rights, duties and liabilities in respect of any expense, liability, loss, claim or proceedings arising from damage or injury to property and related insurance.
6.6	Liability in respect of excepted risks.
6.7	Option A – Rights and duties when insurance of new buildings is the responsibility of the contractor;
	Option B – Duty to insure new buildings as an alternative to Option A; and
	Option C – Duty to insure the works, structures and contents when the works relate to extension or conversion of an existing building.
6.9	Duty to ensure that the joint names policy covers sub-contractors in the event that the employer insures the works under Options B and C.
6.10	Rights, duties and liabilities in respect of the non-availability of terrorism cover.
6.11.3	Right to inspect the contractor's professional indemnity insurance when the contractor undertakes design work in respect of any CDP.
6.12	Right to discuss the means of protection in the event that professional indemnity insurance ceases to be available at commercially reasonable rates.
6.14	Duty to ensure that the employer's personnel comply with the joint fire code.
6.15	Rights duties and liabilities in respect of remedial measures in the event of a breach of the joint fire code.
6.16	Liability for increased costs if, after the base date, the joint fire code is amended.
7.1	Right to assign the contract (with the consent of the contractor).

7.2	Right, after practical completion, to assign the right to bring proceedings in the name of the employer.
7.4	Duty to give notice by one of the three methods prescribed.
7.5	Duty to execute any required warranties in the same manner as the contract.
7A	Rights and duties in respect of third party rights.
7B	Rights, duties and liabilities in respect of rights which may be conferred upon a funder.
7C	Rights and duties in respect of securing collateral warranties in favour of purchasers and/or tenants.
7D	Rights and duties in respect of securing a collateral warranty in favour of a funder.
7E	Rights and duties in respect of securing collateral warranties from sub-contractors in favour of purchasers and/or tenants.
7F	Rights and duties in respect of securing collateral warranties from sub-contractors.
8.2.1	Duty not to terminate the contractor's employment in an unreasonable or vexatious manner.
8.3	Right to retain common law rights and remedies in the event of determination.
8.4	Rights and duties in respect of termination of the contractor's employment in the event of a default by the contractor.
8.5	Optional rights and duties in respect of termination of the contractor's employment in the event of the contractor's bankruptcy or insolvency.
8.6	Right to terminate the contractor's employment in the event of corruption by the contractor.
8.7	Rights and duties in respect of the employment of other persons to carry out and complete the works following termination of the contractor's employment.
8.8	Rights and duties if the employer decides not to proceed with the works following termination of the contractor's employment.
8.9	Liability in the event of the continuance or repetition a specified default or suspension event.
8.10	Duties in the event of bankruptcy or insolvency of the employer.
8.11	Rights, duties and liabilities in the event of determination of the employment of the contractor when neither party is at fault.
8.12	Rights, duties and liabilities in the event of determination of the contractor's employment as a result of the employer's default, bankruptcy or insolvency.
9.1	Right to seek to resolve a dispute through mediation – although to be successful both parties have to want to settle by mediation.
9.2	Rights and duties in respect of adjudication.
9.3 to 9.8	Rights and duties in respect of arbitration.

The contractor

Recitals

Second	Duty to provide a fully priced copy of the bills of quantities and the optional duty to provide a priced activity schedule.
Third	Duty to sign the contract drawings and the contract bills.
Fifth	Right to be provided with an information release schedule if an information release schedule is to be used on the project.
Seventh	Duties, liabilities and rights in respect of any CDP.
Eighth	Right to be provided with documents stating the requirements for the design and construction of the CDP.
Ninth	Duty to provide the employer with 'contractor's proposals' for the design and construction of the contractor's designed portion.
Tenth	Right to rely on the employer having examined the 'contractor's proposals' to establish if they appear to meet the 'employer's requirements'. Duty to sign or initial the 'employer's requirements', 'contractor's proposals' and the CDP analysis.

Articles

1	Duty to carry out and complete the works.
6	Rights, duties and liabilities in respect of CDM Regulation irrespective of whether or not he is the principal contractor.
7	Right to refer a dispute to adjudication.
8	Right to refer a dispute to arbitration if the parties have agreed this method for finally determining disputes.

Clauses

1.9	Right to be issued with all architect's certificates.
2.1	Duty to carry out and complete the works in compliance with the contract documents.
2.2	Duty to carry out and complete the contractor's designed portion in accordance with the contract documents and under the direction of the architect, whilst complying with the CDM Regulations.

2.3.1 Duty to comply with the required standards of materials, goods and workmanship.

2.3.2 Duty to comply with the required standards of materials, goods and workmanship stated in the contract bills for any CDP.

2.3.3 Duty to provide the standards of materials, goods and workmanship appropriate to the works where elements of the CDP are not stated in the contract bills.

2.3.4 Duty to provide proof of compliance of materials and goods and the right to be informed of unsatisfactory work within a reasonable time.

2.3.5 Duty to take reasonable steps to ensure that persons under his control are registered under the Construction Skills Certification Scheme or an equivalent recognised scheme.

2.4 Right to receive possession of the site or, in the case of section work, the relevant part of the site.

2.6.1 Right to consent to the use or occupation of the site or the works, in whole or in part. Duty to ensure that such consent is not unreasonably delayed or withheld.

2.6.2 Right to receive increased insurance premiums, if the contractor insures the works under Option A and the increase in premiums arises as the consequence of the employer's occupation under 2.6.1.

2.7 Right to consent to the employer undertaking work not forming part of the contract when it is not stated in the contract bills. Duty to ensure that such consent is not unreasonably delayed or withheld.

2.8.1 Right to inspect contract documents at all reasonable times.

2.8.2 Right to receive contract documents from the architect/contract administrator.

2.8.3 Duty to keep available contract documents on site for inspection by the architect/contract administrator.

2.8.4 Right to keep confidential any rates or prices in the contract bills, except where required for the purpose of the contract. Duty not to use contract documents for any other purpose than the contract.

2.9 Right to receive from the architect/contract administrator contract documents, drawings and bills of quantities.

2.9.1.1 Right to receive from the architect/contract administrator descriptive schedules and the like.

2.9.1.2 Duty to provide and update the master programme.

2.9.2.1 Duty to provide contractor's design documents.

2.9.2.2 Duty to provide works information (levels etc.) in relation to the contractor's design documents.

2.9.3 Duty to provide contractor's design documents in accordance with the contractor's design submission procedure.

2.10 Rights, duties and liabilities in respect of levels and setting out the works.

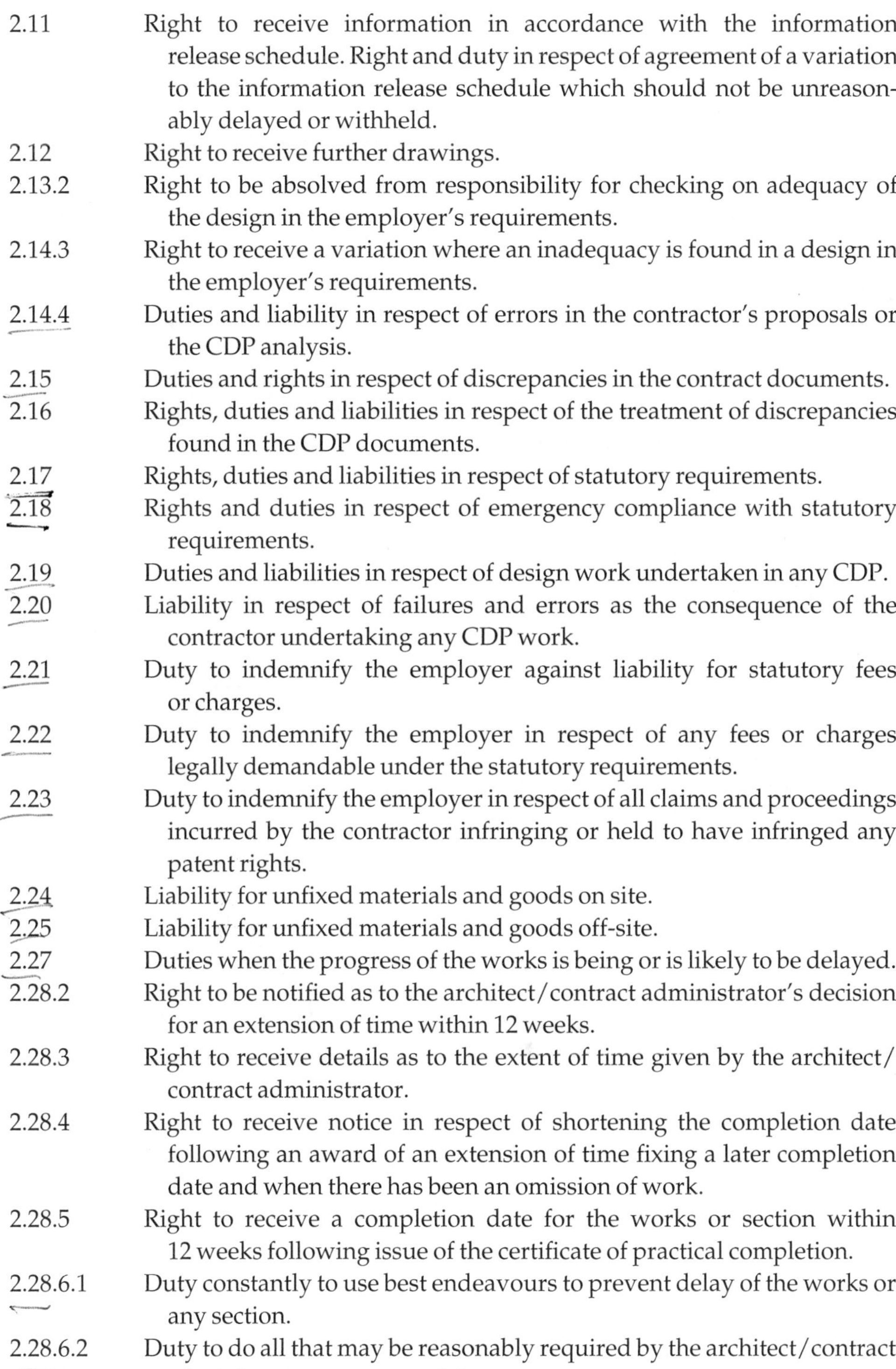

2.11 Right to receive information in accordance with the information release schedule. Right and duty in respect of agreement of a variation to the information release schedule which should not be unreasonably delayed or withheld.

2.12 Right to receive further drawings.

2.13.2 Right to be absolved from responsibility for checking on adequacy of the design in the employer's requirements.

2.14.3 Right to receive a variation where an inadequacy is found in a design in the employer's requirements.

2.14.4 Duties and liability in respect of errors in the contractor's proposals or the CDP analysis.

2.15 Duties and rights in respect of discrepancies in the contract documents.

2.16 Rights, duties and liabilities in respect of the treatment of discrepancies found in the CDP documents.

2.17 Rights, duties and liabilities in respect of statutory requirements.

2.18 Rights and duties in respect of emergency compliance with statutory requirements.

2.19 Duties and liabilities in respect of design work undertaken in any CDP.

2.20 Liability in respect of failures and errors as the consequence of the contractor undertaking any CDP work.

2.21 Duty to indemnify the employer against liability for statutory fees or charges.

2.22 Duty to indemnify the employer in respect of any fees or charges legally demandable under the statutory requirements.

2.23 Duty to indemnify the employer in respect of all claims and proceedings incurred by the contractor infringing or held to have infringed any patent rights.

2.24 Liability for unfixed materials and goods on site.

2.25 Liability for unfixed materials and goods off-site.

2.27 Duties when the progress of the works is being or is likely to be delayed.

2.28.2 Right to be notified as to the architect/contract administrator's decision for an extension of time within 12 weeks.

2.28.3 Right to receive details as to the extent of time given by the architect/contract administrator.

2.28.4 Right to receive notice in respect of shortening the completion date following an award of an extension of time fixing a later completion date and when there has been an omission of work.

2.28.5 Right to receive a completion date for the works or section within 12 weeks following issue of the certificate of practical completion.

2.28.6.1 Duty constantly to use best endeavours to prevent delay of the works or any section.

2.28.6.2 Duty to do all that may be reasonably required by the architect/contract administrator to prevent delay.

2.29	Rights in respect of relevant events.
2.30	Right to have the works or a section certified as being practically complete when in the opinion of the architect/contract administrator they are complete.
2.31	Liability for non-completion of the works or section.
2.32	Rights, duties and liabilities in respect of the payment or allowance of liquidated damages.
2.33 to 2.37	Rights, duties and liabilities in the event of the employer taking partial possession of the works.
2.38	Duty to remedy all shrinkages and defects appearing in the rectification period within a reasonable time and at no extra cost. Liability in respect of the architect/contract administrator instructing that defects, shrinkages and other faults are not to be made good, with an appropriate deduction from the contract sum.
2.40	Duty to provide, without charge, as-built drawings and contractor's design information subject of any CDP.
2.41.1	Rights in respect of any CDP for which the copyright is vested in the contractor.
2.41.2	Right to be paid all the monies due under the contract before the licence to use the contractor's design is given to the employer.
2.41.3	Right not to be held liable for issues arising from the contractor's design, except when it is used for the purpose for which it was prepared.
3.1	Duty to give access to the architect/contract administrator and authorised persons.
3.2	Duty to keep a competent 'person-in-charge' upon the site.
3.3	Right to be notified of any change in the employer's representative.
3.4	Duty to afford the clerk of works every reasonable facility.
3.5	Right to object to the appointment of any person nominated as a replacement architect/contract administrator or quantity surveyor.
3.6	Liability to carry out and complete the works in accordance with the contract.
3.7.1	Duties, rights and liabilities in respect of not sub-letting part of the works.
3.7.2	Duties, rights and liabilities in respect of not sub-letting the design for the CDP.
3.8.1	Right to select sub-contractors from list annexed to the contract bills.
3.8.2	Right, with consent of the employer, to add additional persons to a list contained in the contract bills for work described and contained therein.
3.8.3	Duty, together with the employer, to add names to the list in 3.8.2 when the list falls below three persons.
3.9	Duties and liabilities in respect of conditions which should be applicable in each sub-contract.

3.10	Rights, duties and liabilities in respect of architect/contract administrator's instructions.
3.11	Liability in respect of non-compliance with an architect/contract administrator's instruction.
3.12	Rights and duties in respect of architect/contract administrator's instructions.
3.13	Right to request the architect/contract administrator to state which provision of the conditions empowers him to issue any particular instruction.
3.14.2	Right to make reasonable objection to an architect/contract administrator's instruction.
3.17	Rights, duties and liabilities in respect of instructions for opening up work.
3.18	Rights, duties and liabilities for work not in accordance with the contract.
3.19	Rights, duties and liabilities for workmanship not in accordance with the contract.
3.20	Right to receive reasons for the architect/contract administrator's dissatisfaction with work/workmanship within a reasonable time.
3.22	Duties in respect of antiquities being found during the works.
3.23	Duty to comply with an architect/contract administrator's instruction regarding antiquities.
3.24	Contractor's right to receive reimbursement for loss and/or expense suffered following the discovery of antiquities.
3.25	Rights, duties and liabilities in respect of complying with the CDM Regulations.
3.26	Right to be notified if the employer replaces the planning supervisor or principal contractor.
4.3.1	Right to have the contract sum adjusted by the provisions set out in the clause.
4.3.3	Right to have costs incurred under the prescribed clauses added to the contract sum.
4.5	Duties and rights in respect of final adjustment of the contract sum.
4.6	Rights to be paid VAT properly chargeable in addition to the contract sum.
4.8	Liabilities in respect of advance payments.
4.12	Right to submit an application for interim payment for what he considers to be the gross valuation.
4.13	Rights, duties and liabilities in respect of interim payments due.
4.14	Right to suspend performance of obligations under the contract in respect amounts due.
4.15	Rights, duties and liabilities in respect of final payments due from the employer, or due to the employer, as the case may be.

4.17	Rights and duties in respect of payment for off-site materials or goods.
4.18	Rights in respect of retention.
4.19	Rights, duties and liabilities in respect of the provision of a bond in lieu of retention.
4.23	Rights and duties in respect of the recovery of direct loss and/or expense.
4.24	Right to recover direct loss and/or expense for the prescribed matters.
4.26	Rights to loss and/or expense for matters materially affecting regular progress, which are without prejudice to any other rights that the contractor may possess.
5.1.2	Duty to observe any obligations or restrictions in regard to access to or use of the site, and working space, working hours and the order of work.
5.2.1	Rights and duties with regard to valuing variations.
5.3	Rights and duties in respect of the acceptance of a Schedule 2 quotation.
5.4	Right to be present during the measurement by the quantity surveyor for the purpose of a valuation.
6.1 & 6.2	Rights, duties and liabilities in respect of injury to or the death of persons and loss of or damage or injury to property real or personal and related insurance.
6.4	Duties and liabilities in respect of insuring the contractor's liability.
6.5	Rights, duties and liabilities in respect of any expense, liability loss claim or proceedings or damage or injury to property real or personal and related insurance.
6.6	Right not to indemnify the employer in respect of excepted risks.
6.7	Option A – Duty and liability to insure new buildings in joint names; Option B – Rights and duties to insure new buildings in joint names as an alternative to Option A; and Option C – Rights and duties in respect of the employer insuring the works when they relate to extension or conversion of an existing building.
6.9	Duty to ensure that the joint names policy covers sub-contractors in the event that the contractor insures the works.
6.10	Rights, duties and liabilities in respect of the non-availability of terrorism cover.
6.11	Duty to take out and maintain professional indemnity insurance when the contractor undertakes design work in respect of any CDP.
6.12	Duty to give notice to discuss the best means of protecting the employer in the event that professional indemnity insurance ceases to be available at commercially reasonable rates.
6.14	Duty to ensure that the contractor's personnel comply with the joint fire code.

6.15	Rights duties and liabilities in respect of remedial measures in the event of a breach of the joint fire code.
6.16	Right to have increased costs added to the contract sum if, after the base date, the joint fire code is amended.
7.1	Right to assign the contract (with the consent of the employer).
7.2	Continued liability in the event of assignment of the employer's rights to bring proceedings.
7.5	Duty to execute any required warranties in the same manner as the contract.
7A	Rights and duties in respect of third party rights.
7B	Rights, duties and liabilities in respect of rights which may be conferred upon a funder.
7C	Rights, duties and liabilities in respect of collateral warranties in favour of purchasers and/or tenants.
7D	Rights, duties and liabilities in respect of a collateral warranty in favour of a funder.
7E	Rights and duties in respect of securing collateral warranties from sub-contractors in favour of purchasers and/or tenants.
7F	Rights and duties in respect of securing collateral warranties from sub-contractors for the employer.
8.2	Right not to have employment terminated unreasonably or vexatiously.
8.3	Liability for any other rights or remedies available to the employer outside of the contract.
8.4	Rights, duties and liabilities in respect of termination of the contractor's employment.
8.5	Rights, duties and liabilities in respect of termination in the event of the contractor's bankruptcy or insolvency.
8.6	Duties and liabilities in respect of corruption.
8.7	Rights, duties and liabilities in respect of termination of the contractor's employment.
8.8	Rights and duties if the employer decides not to proceed with the works following termination of the contractor's employment.
8.9	Rights and duties in respect in of an employer default.
8.10	Duties in the event of bankruptcy or insolvency of the employer.
8.11	Rights, duties and liabilities in the event of determination of the employment of the contractor when neither party is at fault.
8.12	Rights, duties and liabilities in the event of determination of the contractor's employment as a result of the employer's default, bankruptcy or insolvency.
9.1	Right to seek to resolve a dispute through mediation – although to be successful both parties have to want to settle by mediation.
9.2	Rights and duties in respect of adjudication.
9.3 to 9.8	Rights and duties in respect of arbitration.

The architect/contract administrator

Clauses

1.9	Duty to issue certificates to the employer and simultaneously to the contractor.
2.3.1	Right to decide whether consent is given to substitute materials.
2.3.3	Right to have work/workmanship to his reasonable satisfaction.
2.3.4	Right to request reasonable proof that goods and materials are in accordance with the contract documents.
2.8.2	Duty to provide contract documents following execution of the contract.
2.8.3	Right to have contract documents and information made available by the contractor.
2.8.4	Duty not to divulge any rates or prices in the contract bills except for the purpose of the contract.
2.9	Rights and duties in respect of the issue and supply of contract documents, the contractor's design submission procedure and the master programme.
2.10	Duty to determine the levels required for execution of the works. Right to issue an instruction that errors made by the contractor in setting out the works remain.
2.11	Rights and duties in respect of the information release schedule.
2.12	Duties and rights in respect of the issue of drawings from time to time that may reasonably be necessary for the contractor to complete the works.
2.15	Duty to issue instructions in regard to discrepancies in or divergences between documents.
2.16.1	Right not to be obliged to issue instructions in respect of any CDP.
2.16.2	Rights and duties in respect of discrepancies in the employer's requirements.
2.17	Rights and duties in respect of divergences in the statutory requirements.
2.18	Rights and duties in respect of emergency compliance with the statutory requirements.
2.27	Right to be notified in respect of delay or likely delay to the progress of the works or section.
2.28	Rights and duties in respect of fixing a later completion date due to a relevant event.
2.30	Duty to issue practical completion certificates when, in his opinion, the works or a section is practically complete.
2.31	Duty to issue a non-completion certificate if the contractor fails to complete the works or section within the relevant completion date.
2.33	Duty to issue a written statement in respect of the employer taking partial possession of the works or section.

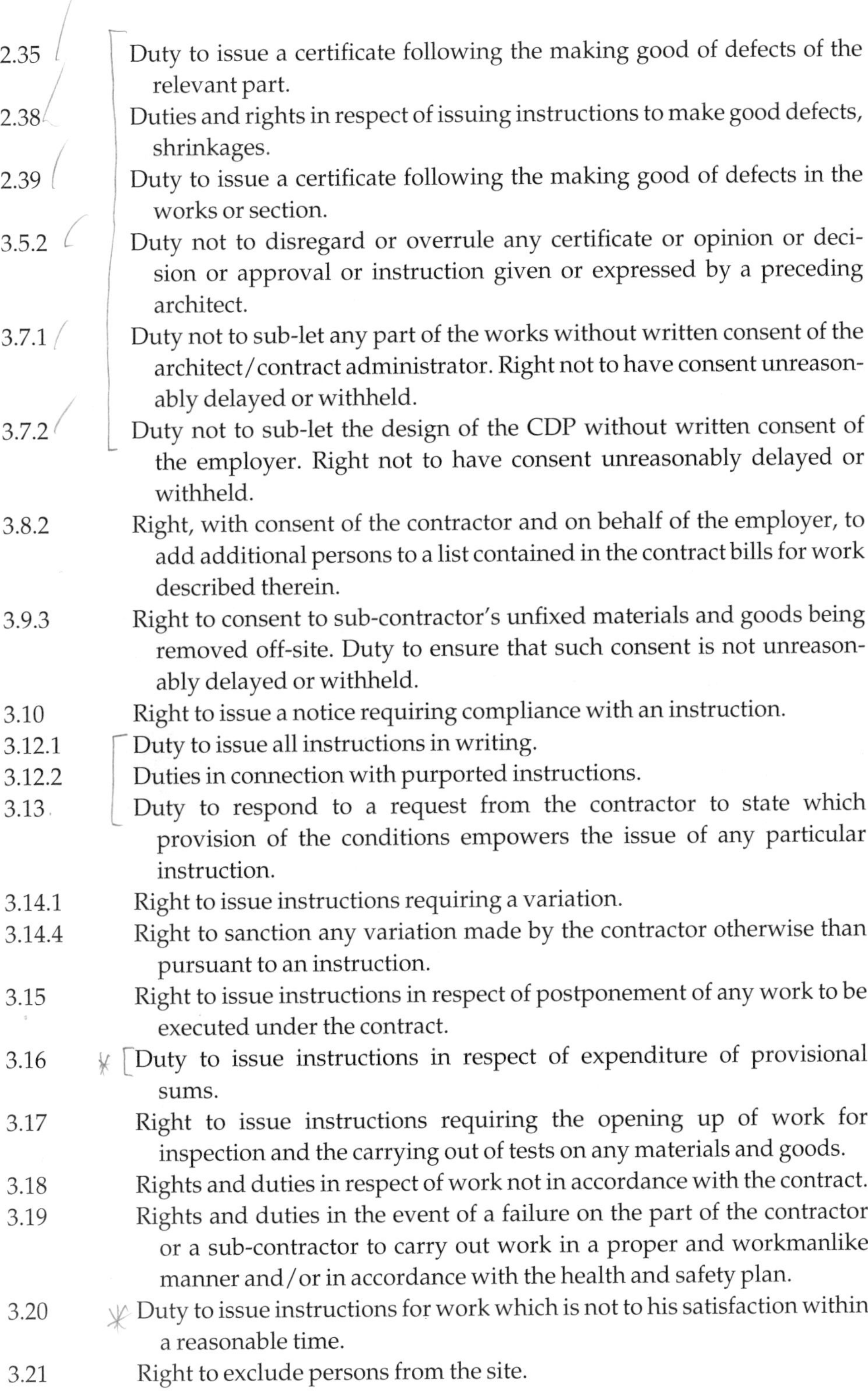

2.35 Duty to issue a certificate following the making good of defects of the relevant part.

2.38 Duties and rights in respect of issuing instructions to make good defects, shrinkages.

2.39 Duty to issue a certificate following the making good of defects in the works or section.

3.5.2 Duty not to disregard or overrule any certificate or opinion or decision or approval or instruction given or expressed by a preceding architect.

3.7.1 Duty not to sub-let any part of the works without written consent of the architect/contract administrator. Right not to have consent unreasonably delayed or withheld.

3.7.2 Duty not to sub-let the design of the CDP without written consent of the employer. Right not to have consent unreasonably delayed or withheld.

3.8.2 Right, with consent of the contractor and on behalf of the employer, to add additional persons to a list contained in the contract bills for work described therein.

3.9.3 Right to consent to sub-contractor's unfixed materials and goods being removed off-site. Duty to ensure that such consent is not unreasonably delayed or withheld.

3.10 Right to issue a notice requiring compliance with an instruction.

3.12.1 Duty to issue all instructions in writing.

3.12.2 Duties in connection with purported instructions.

3.13 Duty to respond to a request from the contractor to state which provision of the conditions empowers the issue of any particular instruction.

3.14.1 Right to issue instructions requiring a variation.

3.14.4 Right to sanction any variation made by the contractor otherwise than pursuant to an instruction.

3.15 Right to issue instructions in respect of postponement of any work to be executed under the contract.

3.16 Duty to issue instructions in respect of expenditure of provisional sums.

3.17 Right to issue instructions requiring the opening up of work for inspection and the carrying out of tests on any materials and goods.

3.18 Rights and duties in respect of work not in accordance with the contract.

3.19 Rights and duties in the event of a failure on the part of the contractor or a sub-contractor to carry out work in a proper and workmanlike manner and/or in accordance with the health and safety plan.

3.20 Duty to issue instructions for work which is not to his satisfaction within a reasonable time.

3.21 Right to exclude persons from the site.

3.22.3 Right to be informed of the discovery of antiquities or other objects of interest or value.

3.23 Duty to issue instructions in the event of the discovery of antiquities or other objects of interest or value.

3.24 If, in his opinion, the contractor has suffered loss and/or expense, the architect/contract administrator has a duty to ascertain, or instruct the quantity surveyor to ascertain, the amount.

3.25.2 Right to be notified by the employer of any amendments by the contractor to the health and safety plan.

4.3 Duties in respect of adjusting the contract sum.

4.5.1 Right to receive all documents necessary for the purposes of adjusting the contract sum.

4.5.2 Duty to ascertain the amount of any loss and/or expense incurred by the contractor if the quantity surveyor is not instructed to do so.

Duty to send the contractor a copy of the statement which adjusts the contract sum and states the amount ascertained for any contractor's loss and/or expense under clauses 3.24 and 4.23.

4.9 Rights and duties in respect of interim certificates.

4.14 Right to receive a copy of the contractor's notice of intention to suspend performance.

4.15.1 Duties in respect of the final certificate.

4.17 Rights in respect of payments for materials off-site.

4.18.2 Duty to prepare, or instruct the quantity surveyor to prepare, a statement of retention.

4.19 Rights and duties in respect of the retention bond when applicable.

4.21 Option A – Rights in respect of contribution, levy and tax fluctuations.

Option B – Rights in respect of labour and materials cost and tax fluctuations.

Option C – Duties in respect of fluctuations calculated by use of price adjustment formulae.

4.23 Rights and duties when the contractor has incurred or is likely to incur direct loss and/or expense due to deferment of being given possession of the site or because the regular progress of the works has been materially affected by a listed matter.

5.3.1 Rights in respect of issuing an instruction requesting a Schedule 2 quotation.

5.3.2 Right to instruct a variation when the contractor disagrees with the application of a Schedule 2 quotation.

5.7 Right to receive verification in respect of daywork.

6.4.2 Right to receive proof of cover in respect of the contractor's insurance of his liability.

6.5.1 Right to instruct the contractor to take out a policy of insurance in respect of liability, etc. of the employer and to receive the policies and premium receipts therefore.

6.5.2 Right to receive documentary evidence of insurance and duty to deposit it with the employer.

6.7 Right to receive notice of loss or damage to executed work or site materials caused by any risk covered under the joint names policy.

6.11.3 Right to receive documentary evidence that professional indemnity insurance is being maintained.

6.14 Duty to comply with the joint fire code where the contract particulars state that the joint fire code applies.

6.15.1 Rights and duties in respect of remedial measures if there is a breach of the joint fire code.

8.4.1 Right to give notice to the contractor in respect of a specified default or defaults.

8.7.2.1 Rights in relation to the removal of temporary buildings, plant, tools, equipment, goods and materials from the works.

8.7.2.3 Right to request, on behalf of the employer, the assignment from the contractor to the employer of the benefit of any agreement.

8.7.4 Duty, in the event of termination of the contractor's employment, in respect of issuing a further certificate to the contractor identifying the cost of the works, payments made to the contractor and any amount of loss and/or expense incurred by the employer for which the contractor is liable and the total amount that would have been payable if the works had been in accordance with the contract.

The quantity surveyor

Clauses

2.8.4 Duties in relation to the confidential nature of the contract documents.

3.24 Duty, if so instructed by the architect/contract administrator, to ascertain the amount of loss and/or expense incurred by the contractor in the event of fossils, antiquities and other objects of interest or value being found on site.

4.5 Rights and duties, if so instructed, in respect of the final adjustment of the contract sum.

4.11 Duty to make interim valuations whenever the architect considers them to be necessary for the purpose of interim certificates. (When clause 4.10 applies this duty is not at the discretion of the architect – a valuation is then obligatory prior to each interim certificate).

4.12 Rights and duties in the event of the contractor submitting an application prior to an interim certificate.

4.18.2 Duty, if so instructed by the architect/contract administrator, to prepare a statement of retention.

4.21 Option A – Rights in respect of contribution, levy and tax fluctuations.
Option B – Rights in respect of labour and materials cost and tax fluctuations.
Option C – Rights in respect of fluctuations calculated by use of price adjustment formulae.

4.23 Duty, if so instructed by the architect, to ascertain the direct loss and/or expense incurred or likely to be incurred by the contractor due to deferment of being given possession of the site or because the regular progress of the works has been materially affected by a listed matter.

4.23.3 Right to receive, from the contractor, details of loss and/or expense (if requested).

5.2.1 Duty to value variations when not agreed by the employer and contractor.

5.3 Right to receive, from the contractor, Schedule 2 quotations. Duty in respect of valuing variations to work for which a Schedule 2 quotation has been accepted.

5.4 Duty to give the contractor the opportunity of being present and of taking notes at the time of taking measurements for the valuation of variations.

The planning supervisor

Clause

3.25.2 Right to be notified by the employer of any amendments by the contractor to the health and safety plan.

3.25.3 Rights in respect of the information necessary for the preparation of the health and safety file.

The clerk of works

Clause

3.4 Duty to act solely as inspector on behalf of the employer under the direction of the architect.
Right to give directions to the person-in-charge.

3.22.3 Right to be informed by the contractor of the discovery on site and the precise location of fossils, antiquities and other objects of interest or value.

The person-in-charge

Clause

3.2 Duty to receive instructions of the architect/contract administrator and directions of the clerk of works.

Conclusion

One may have gleaned from the foregoing rights, duties and liabilities that where one party has a particular right there is, more often than not, a reciprocal duty on the other party. It is the recording of decisions in respect of these rights and duties which might ultimately decide whether the project is a success or a failure. Over many decades the JCT, through the cross-section of bodies representing the industry, has tried to define clearly and balance properly the rights and duties between employer and contractor and, by and large, be fair to both parties.

Chapter 24
Meetings

Initial meeting

As soon as practicable after the contract has been placed, the building team should meet. Although this initial meeting may take place on site, it will more probably take place in the offices of the project manager, the architect/contract administrator or the main contractor.

The manner in which the initial meeting is conducted will greatly influence the success of the project, and succinct clear direction from the chair will be a strong inducement to a similar response from others. Since at this stage the person having the most complete picture of the project is likely to be the project manager, if there is one, or the architect/contract administrator, if there is not, it is logical that the project manager or the architect/contract administrator should take the chair. The arrangements for this meeting should be discussed with the main contractor beforehand. It is suggested that as many of the following as are involved with the project should be requested to attend:

- employer
- project manager
- planning supervisor
- architect/contract administrator
- quantity surveyor
- structural engineer
- building services engineer(s)
- contractor
- clerk of works.

It is also suggested that the agenda for the initial meeting should include the following matters:

- introductions
- factors affecting the carrying out of the works

- programme
- sub-contractors and suppliers
- lines of communication
- bonds, collateral warranties and insurances (see Chapter 22)
- financial matters
- procedure to be followed at subsequent meetings.

Introductions

The introduction of those attending needs no elaboration, although it is more than just a formality as it establishes the initial contact between individuals who must work together in harmony if the project is to run smoothly.

Factors affecting the carrying out of the works

These should be described fully in the contract documents but may require emphasis and clarification at the initial meeting. Such factors may include:

- existing mains/services
- site investigation, including soils and ground water information
- access to the site
- parking restrictions
- use of the site
- risks to health and safety and health and safety plans
- work outside the site boundary
- concurrent work by others
- named listed sub-contractors
- equivalent products
- Considerate Constructors Scheme
- photographic records
- prohibited products
- protection of products
- samples
- building lines, setting out, critical dimensions and tolerances
- co-ordination of engineering services
- quality control
- use of explosives and pesticides and on-site burning of rubbish
- fire prevention
- protection of the works, existing retained features, adjoining buildings, work people and the public
- working hours and overtime working
- location of temporary accommodation, spoil heaps, etc.

- temporary fences, hoardings, screens, roads, name boards, etc.
- temporary services and facilities.

Programme

The SBC in clause 2.9.1.2 requires that the contractor shall provide the architect/contract administrator with two copies of the master programme for the execution of the works as soon as possible after the execution of the contract.

In addition the contractor is also required to update it within 14 days of any decision by the architect/contract administrator that creates a new completion date. In any event the master programme should be reviewed regularly and updated whenever necessary. It is to be noted that the master programme is not a contract document and SBC clause 2.9.1 makes it clear that nothing contained in the master programme can impose any obligation on the contractor beyond those obligations imposed by the contract documents. Some forms of contract do not require the contractor to provide a programme. In such circumstances if one is wanted the need should be stated in the bills of quantities, specification or the contract documents themselves.

In *Glenlion Construction Limited* v. *The Guinness Trust* it was decided that:

- The contractor is entitled to complete the works on or before the date for completion stated in the contract particulars, or any later completion date fixed pursuant to the appropriate conditions of contract; and
- the contractor is not entitled to expect the design team to provide design information to enable him to complete the works before the date for completion stated in the contract particulars, or any later completion date fixed pursuant to the appropriate conditions of contract.

At first sight these two statements may appear to conflict with each other but with further thought it will be apparent that they do not. This is because even if the contractor is capable of completing the works before the date for completion, it may not be possible for the design team to provide design information sooner than would be necessary for the contract completion date.

Prior to the initial meeting the contractor should determine, as far as possible, major material delivery dates and required construction periods from sub-contractors. From this information and the Information Release Schedule, when provided by the employer (see SBC fifth recital and clause 2.11), the contractor should formulate a draft master programme for the works which should then be circulated to all members of the building team. Whether or not an information release schedule has been provided, it is quite usual for the contractor to indicate on the draft master programme the latest dates by which drawings, schedules, instructions for placing orders and other information are required from the design team.

The draft master programme should be considered and, if necessary, adjustments agreed during the initial meeting. Following this the contractor should prepare his definitive master programme and circulate it to all concerned. A simple bar chart programme is given as Example 24.1 at the end of this chapter.

The importance of adhering to dates once agreed, whether they be in respect of materials deliveries, execution of the work or the production of further necessary information by the design team, cannot be over-stressed.

Sub-contractors and suppliers

Many contractual problems arise as a result of the contractor failing to issue instructions in relation to sub-contractors and suppliers timeously and as a result of poor communications between the main contractor and those firms engaged. It is therefore advisable at the initial meeting for the architect/contract administrator to inform the main contractor of his intended instructions that might affect sub-contractors and suppliers. It is also advisable to make it clear to the main contractor that the sub-contractors and suppliers are contractually his responsibility.

It is essential, for the smooth and efficient running of the project, that all the sub-contractors are engaged sufficiently early for the work concerned to be integrated into the contractor's programme of work without causing disruption. It is also essential that all members of the building team allow sufficient time in their work schedule to carry out properly the various complex and often time-consuming procedures involved in the selection and appointment of sub-contractors and suppliers. Following the correct procedures and providing the appropriate documentation when appointing sub-contractors is of the utmost importance and is dealt with in Chapter 20.

Lines of communication

It is important at the initial meeting for the chairman to emphasise the need for all members of the building team to abide by the formal lines of communication provided for in the SBC and to make reference to the golden rules of communication given in Chapter 1.

Financial matters

It is desirable at the initial meeting for the quantity surveyor to run through the various procedures set down in the SBC relating to financial matters. The sheer number of clauses relating to financial matters might mean that the quantity surveyor has to limit the discussion on procedures to those regarding interim valuations, valuation of variations and payment. This can, however, be tailored to the employer and the contractor's experience and the specifics of the project. Clauses concerning financial matters and which might require a degree of explanation comprise:

2.21 Fees or charges legally demandable
2.22 Royalties and patent rights – Contractor's indemnity
2.23 Patent rights – Instructions
2.24 Materials and goods – on-site
2.25 Materials and good – off-site
2.32 Payment or allowance of liquidated damages
2.36 Partial Possession by Employer – Insurance – Relevant Part
2.37 Partial Possession by Employer – Liquidated Damages – Relevant Part
3.9 Conditions of sub-letting
3.24 Antiquities – Loss and expense arising
4.1 Work included in Contract Sum
4.2 Adjustment only under the Conditions
4.3 Items included in adjustments
4.4 Taking adjustments into account
4.5 Final adjustment
4.6 Value added tax
4.7 Construction Industry Scheme (CIS)
4.8 Advance payment
4.9 Issue of Interim Certificates
4.10 Amounts due in Interim Certificates
4.11 Interim valuations
4.12 Application by Contractor
4.13 Interim Certificates – payment
4.14 Contractor's right of suspension
4.15 Final Certificate – issue and payment
4.16 Ascertainment (Gross Valuation)
4.17 Off-site materials and goods
4.18 Rules on treatment of Retention
4.19 Retention Bond
4.20 Retention – rules for ascertainment
4.21 Choice of fluctuation provisions
4.22 Non-applicability of a Schedule 2 Quotation
4.23 Loss and Expense – Matters materially affecting the regular progress
4.24 Relevant Matters
4.25 Amounts ascertained – addition to Contract Sum
4.26 Reservation of Contractor's rights and remedies
5.2 Valuation of Variations and provisional sum work
5.3 Schedule 2 Quotation
5.4 Contractor's right to be present at measurement
5.5 Giving effect to Valuations, agreements etc.

5.6 The Valuation Rules – Measurable Work
5.7 The Valuation Rules – Daywork
5.8 The Valuation Rules – Contractor's Designed Portion – Valuation
5.9 The Valuation Rules – Change of conditions for other work
5.10 The Valuation Rules – Additional provisions
6.4 Insurance against personal injury and property damage – Contractor's insurance liability
6.5 Insurance against personal injury and property damage – Contractor's insurance of employer's liability
6.7 Insurance of the Works
6.11 CDP Professional Indemnity Insurance – Obligation to insure
6.12 CDP Professional Indemnity Insurance – Increased cost and non-availability
8.4 Termination by Employer – Default by Contractor
8.5 Termination by Employer – Insolvency of Contractor
8.6 Termination by Employer – Corruption
8.7 Termination by Employer – Consequences of termination under Clauses 8.4 to 8.6
8.8 Termination by Employer – Employer's decision not to complete the Works
8.9 Termination by Contractor – Default by Employer
8.10 Termination by Contractor – Insolvency of Employer
8.11 Termination by either Party
8.12 Consequences of termination under Clauses 8.9 to 8.11, etc.

It is worth stressing that no adjustment in respect of varied work will be made to the contract sum unless the matter is covered by an architect's instruction issued in accordance with the terms of the contract. Also, it is worth reminding the contractor of any other financial matters required by the contract documents, such as the provision by him of a cash flow forecast showing the gross valuation of the works at the date of each interim certificate.

Procedure to be followed at subsequent meetings

Formal site meetings are not to be confused with the architect's periodic site visits and the numerous meetings between the main contractor and others that are necessary to the progress of the works. The frequency of formal site meetings will vary with the size and complexity of the project, the stage of the job and any difficulties being encountered. It would be unusual for such meetings to be held less frequently than once every four weeks.

The requirements of the employer as to the nature, normal frequency, procedures and location for formal site meetings should be given in the contract documents. These meetings will be chaired by the project manager, if there is one, or the architect/contract administrator, if there is not, and will normally be attended by the design team, the main contractor and such sub-contractors as are requested to be present.

The object of the formal site meetings is to provide a forum for the presentation by the contractor of his progress report which should include at least:

- a progress statement by reference to the master programme for the works
- details of any matters materially affecting the regular progress of the works
- any requirements for further drawings, details and instructions.

It also provides an opportunity for contractors to raise any queries they may have with the design team as a whole and to report formally on health and safety issues.

The project manager or architect/contract administrator should notify the rest of the design team and the main contractor of the dates and times for site meetings and should request those whose presence is required to attend. It should be the responsibility of the main contractor to invite to the site meetings representatives of those sub-contractors and suppliers whom he or the design team wish to be present. Due consideration should be given to the value of people's time, with care being taken not to invite to meetings those whose presence is not really necessary.

... a great aid to the smooth running of the project

The project manager or architect/contract administrator and the main contractor should agree the agenda for each site meeting in advance. A standard form of agenda is useful as a model and an *aide memoire*. A typical site meeting agenda is given as Example 24.2 at the end of this chapter. The minutes of site meetings will normally be taken and distributed by the chairman of the meeting. They must be impartial and concise, and accurately record all decisions reached and actions required. Typical site meeting minutes are given as Example 24.3.

It is worthy of note that a badly run site meeting can do untold harm and be a serious waste of time for all attending whereas a well-managed meeting can be a great aid to the smooth running of the project. Further, there is little doubt that well-run meetings maintain the team's interest in the project and impart a sense of involvement and urgency which is difficult to achieve by any other means.

Contractor's meetings

The main contractor will normally hold meetings with sub-contractors and suppliers shortly before the formal site meetings. Such meetings help facilitate the accurate reporting of progress and ensure that all sub-contractors and suppliers have their information requirements noted and met. They also provide an opportunity to establish requirements for holes, chases, recesses, fixings and the like before work is put in hand and thereby avoid conflict with other work.

Employer's meetings

The employer may also hold separate meetings with the rest of the design team during the progress of the works. Such meetings are normally used to discuss general progress and to resolve any outstanding design issues, such as colour schemes. They also provide an opportunity to consider the possible effects of proposed variations on the works and on the anticipated final costs.

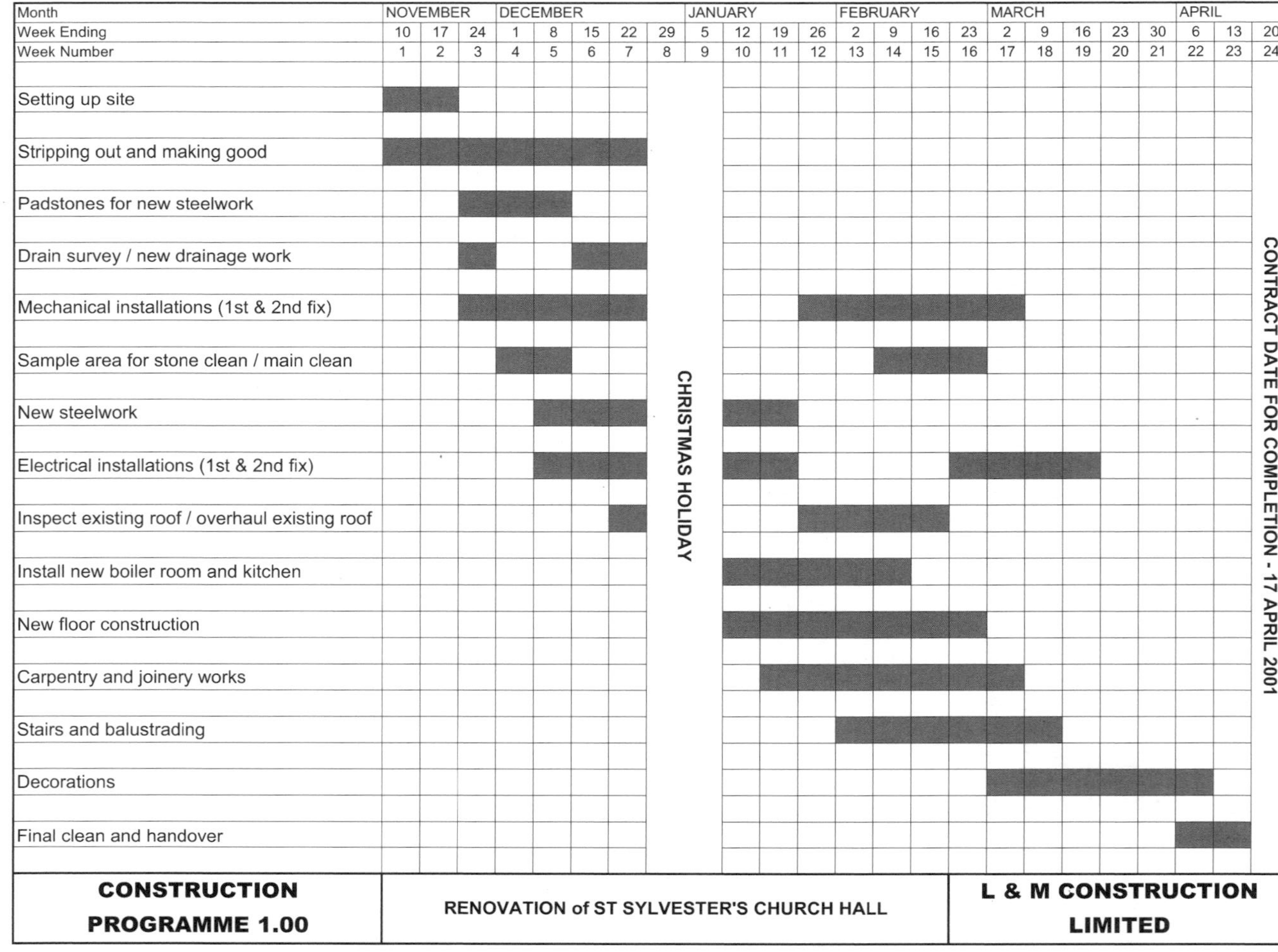

Example 24.1 Construction programme.

Project:	Shops & Offices, Newbridge Street, Borchester
Project ref:	456
AGENDA FOR SITE MEETING	
Date:	7 October 2002 at 10.00am
1.0	Apologies
2.0	Minutes of last meeting
3.0	Contractor's report General report Progress statement by reference to the master programme for the works Details of any matters materially affecting regular progress of the works Information received since last meeting Requirements for further drawings, details and instructions Health and safety matters
4.0	Clerk of Works' report Site matters Quality control Lost time
5.0	Design Consultants' reports Architect Structural Engineer Building Services Engineer
6.0	Quantity Surveyor's report
7.0	Contract completion date Assess likely delays Review factors from previous meeting List factors for review at next meeting Record anticipated completion date
8.0	Any other business
9.0	Date and time of future meetings Site meetings Site visits

Distribution:

Copies	2 Employer	3 Main Contractor	1 Project Manager
	1 Planning Supervisor	1 Architect	1 Quantity Surveyor
	1 Structural Engineer	1 Services Engineer	1 Clerk of Works

Example 24.2 Typical site meeting agenda.

Project:	Shops & Offices, Newbridge Street, Borchester	Ivor Barch Associates Prospects Drive Fairbridge
Project ref:	456	

SITE MEETING NO 4

Date:	7 October 2002	
Location:	Site	
Present:	S Gilbert	Employer (Aqua Products plc)
	A Morley	Main Contractor (L & M Construction Ltd)
	G Mackay	Main Contractor (L & M Construction Ltd)
	B Hunt	Project Manager
	Ivor Barch	Architect (Ivor Barch Associates)
	R W Pipe	Quantity Surveyor (Fussedon Knowles & Partners)
	I Tegan	Structural Engineer (GFP & Partners)
	F Adams	Services Engineer (Black & Associates)
	H Hemmings	Clerk of Works

Item		Action
1.0	<u>Apologies</u>	
	None.	
2.0	<u>Minutes of last meeting</u>	
	Agreed as correct.	
3.0	<u>Contractor's report</u>	
3.1	Progress is generally satisfactory.	
3.2	Still one week behind master programme due to late delivery of bricks.	
3.3	AI5 received and actioned.	
3.4	Details of ironmongery revisions required in next two weeks.	Architect
3.5	No accidents to report.	
4.0	<u>Clerk of Works' report</u>	
4.1	Concern expressed about poor stacking of bricks.	L&MC
5.00 – 8.00	*Continue through agenda*	
9.0	<u>Future meetings</u>	
	Site Meeting – 4 November 2002 at 10.00am.	All
	Architect's site visit – 21 October 2002 at 10.00am.	

Distribution:

Copies	2 Employer	3 Main Contractor	1 Project Manager
	1 Planning Supervisor	1 Architect	1 Quantity Surveyor
	1 Structural Engineer	1 Services Engineer	1 Clerk of Works

Example 24.3 Typical site meeting minutes.

Chapter 25
Site Duties

The architect on site

The development of drawings, models and documentation made it possible in the post-medieval world for the designers of buildings to be able to leave the site environment, to be based in a studio and have a more distant relationship with their creations – and their builders. This meant that the architect became an independent entity, and lost his dual role as architect and master builder combined.

The result was that from time to time, depending on the accuracy of the architects' drawings and the skill of the builders, the designers had to visit their sites to check on the progress of the works and provide interpretation, guidance and leadership as the job proceeded. Clients looked to the architect, not the builder, as their representative and the person responsible for successful outcomes.

Professional services are now provided in an age of litigation. Clients endeavour on an unprecedented scale to pass project risks to their consultants and contractors. The amount of Professional Indemnity insurance held by an architect is checked by potential clients in as much detail as his creative and technical skills. It is not surprising, in this brittle and legally constrained climate, that the definition of site visits by architects has to be worded precisely so as not to imply a depth and rigour of scrutiny that would be either uneconomic or physically impossible for them to provide, and which would open them to legal action for errors or omissions beyond their control.

An architect's site duties vary from those that apply during the feasibility and design stages, which may comprise both subjective and objective observations, complying with the requirements of the project, to the highly formal site visits as prescribed by the form of contract between employer and contractor.

Every participant in the sponsorship, design, project management and construction process tries hard to place risk in the hands of others, but it has to stop somewhere. Architects have had to describe their on-site services with accuracy to avoid being burdened with more unnecessary and unwelcome responsibilities than is reasonable as these carry expensive risk.

The Standard Form of Agreement (SFA/99, updated 2004) says that the architect should make visits to the Works in connection with the architect's design (SFA/99/k/1). The Agreement will stipulate the number of visits to site. Site visits are rarely short or convenient. Should more site visits be considered necessary by the architect and the rest of the design team (for example in works to an existing building) then the client should be asked for fees for additional visits.

Architects visit sites for three main reasons:

- checking that the quality of completed work conforms to agreed quality standards
- checking that the contractor's progress conforms to the programme
- checking that designs are being completed according to plan.

Supervision and inspection

It is important to distinguish between supervision and inspection duties.

Supervision is the responsibility of the main contractor who is bound by the conditions of the SBC to complete the work in a certain time and to a specified standard. Other forms of contract may also provide or imply contractor supervision, for example a design and build arrangement.

Architects, under the terms of their employment, will normally be required to visit the site, at intervals appropriate to the stage of construction, to inspect the progress and quality of the works and to determine that they are being executed in accordance with the contract documents. While on site, it is often possible for them to make sure that problems, which are bound to arise even on the smallest project, are satisfactorily resolved. However, architects are not normally required to make frequent or constant inspections. The term 'inspection' has far reaching implications, implying a depth and rigour of examination to which many architects would not wish to be committed, and for which they might be open to legal recourse should any problems arise which arguably should have been discovered during 'inspections'. The term 'observation' is less onerous. Some bespoke forms of agreement and contract describe the architect as having to 'observe' the work on site. 'Observe' is therefore the term used in conjunction with the activities of the design team throughout the remainder of this chapter.

The nature and extent of supervision arrangements will depend on the size and complexity of the works. On a small project, for example, the contractor may be able to rely on a competent general foreman, while on large projects, if the job is to be properly supervised, a number of foremen and assistants may be required, all working under a site agent or a contracts manager.

Similarly, the architect on a small project may be able to undertake his normal duties by periodic visits; however on large and complex projects, one or more clerks of works may be needed, with possibly a resident architect. In this chapter it is assumed that there will be a clerk of works acting pursuant to SBC clause 3.4, 'as

possibly a resident architect

inspector on behalf of the Employer under the directions of the Architect/Contract Administrator'.

The role of clerk of works may vary slightly depending on whether the appointment is made by the employer or the architect. If appointed by the employer, the role may approach that of a project manager, looking after the employer's broader interests.

Routine site visits

The architect will normally have more time to observe the work when making routine site visits than on those occasions when a formal site meeting is held. The frequency of such visits and the depth of observation will depend on the size and complexity of the work and the speed of the progress being made.

Architects and their representatives are given right of access to sites under most forms of contract, as well as places where work is being fabricated by contractors off-site.

As a matter of safety and courtesy, the architect should make his presence on site known, preferably by prior arrangement, to the clerk of works and the site agent. The latter will normally accompany the visiting architect around the job. The architect should never give instructions to a workman directly. The site agent or general foreman employed by the main contractor is usually the only person designated as 'the competent person-in-charge' under the SBC to receive and act upon architect's instructions. Any oral instructions given during a site visit should be confirmed in writing (see Chapter 26).

When making a site visit to observe the works it is easy to be distracted and to overlook items that require attention. It is a good plan, therefore, for the architect to list in advance particular points to be looked at and any special reason for doing so. For this purpose a standard checklist is a helpful *aide memoire*. Provided that the contents of such a list are of a general nature, they can then be amplified or adapted for each job according to the type of work and form of construction involved. A typical standard checklist is given as Example 25.1 at the end of this chapter.

It is important that accurate and full records are kept of all site visits, together with the results of any tests ordered or completed. One of the key reasons an architect visits a site is to observe and check the quality of the contractors' work against the agreed method statement. Carefully structured reports should contain site photographs, notes and sketches.

Consultants' site visits

In addition to the architect, the structural engineer and services consultants will also make regular site visits to observe the works.

Depending upon the nature of the work, it is often necessary, in the early stages of a project, for the structural engineer to confirm or adjust foundation designs following excavation and to advise on works necessary to stabilise existing buildings. Subsequently, the structural engineer's main concern is in the testing of concrete and other structural elements (see *Samples and testing* below).

Once the services installations are under way, the services consultants are likely to undertake site visits to ensure that the works are meeting the specified requirements.

Inspections by statutory officials

Visits by statutory officials are to be anticipated on any project – with a wider range visiting the larger projects.

Foremost is the building control officer from the local authority. The contractor is required to give notice of particular critical points within the construction process, whereupon a building control officer will visit and advise/comment upon matters pertinent to the building regulations, for example, foundations, drainage, structure, fire protection, etc. It is common for such an inspector to issue directions, but the contractor must, before acting upon them, refer any such direction to the architect for clarification and confirmation as a formal 'instruction'. The architect may wish to consider the inspector's directions for several reasons:

- they may compromise functionality
- they may have an effect on the programme
- they may have cost implications

- there may be preferable, alternative solutions
- there may be the option to discuss with the building control officer whether or not they are really necessary.

A local authority environmental health officer may wish to inspect the site on the grounds of noise or dust pollution.

An officer from the Health and Safety Inspectorate (HSI) is likely to visit the site to inspect such things as site access, general working conditions and scaffolding to ensure that they meet relevant safety criteria. The HSI inspector may also ask to see the contractor's Risk Register.

Subject to the type of building, a Fire and Civil Defence Authority officer may visit to ensure that the means of escape are acceptable, alarms are audible, fire doors adequate, etc.

In addition to such statutory officials, it may be that others, for example NHBC inspectors, surveyors acting for adjoining owners or for insurers and or the employer's financial backers, seek inspections.

Whoever visits the site, it is important that the contractor verifies the purpose of their visit and checks whether it is advisable for any other party to accompany them. The contractor should always record the visit and make a brief report of the reason for it and the outcome.

Records and reports

Where a clerk of works is employed he should be required to keep a daily diary of all matters affecting the project. This diary, to be handed to the architect at the end of the project, should in the interim form the basis of weekly reports to the architect. A standard form of report is available from the Institute of Clerks of Works. A completed copy is given as Example 25.2 at the end of this chapter.

Such reports cover:

- number of men employed daily, in the various trades
- weather reports and particulars of time lost due to adverse weather conditions
- length of any work stoppages
- visitors to site including statutory inspectors
- delays including action taken to remedy defective work
- site directions issued
- drawings/information received on site
- drawings/information required on site
- plant/materials delivered to or removed from site
- general progress in relation to the programme
- general report including a summary of work proceeding
- comments on site conditions/cleanliness/health and safety.

... adverse weather conditions ...

The clerk of works' diary and weekly reports constitute a most useful record of events, which may be referred to should disputes subsequently arise. In the absence of a clerk of works, the contractor should be required to compile a similar factual record of conditions and activities on site and the design team should maintain their own records of site observations including plant, labour, activity and progress, to augment or amplify the contractor's records.

A copy of the contractor's programme should be kept in the office of the clerk of works and the actual progress should be checked and recorded against this programme every week.

It is the job of the clerk of works to keep records of any departures from the production information so that, on completion, architects have all the information necessary to enable them to issue to the employer an accurate set of 'as-built' drawings of the finished building. These records are particularly important where the work is concealed, as for example with foundations, the depth of which may vary from that shown on the original production information.

Progress photographs of the work also form valuable records if taken regularly. This is probably best arranged in conjunction with the contractor so that the whole building team can have the benefit of them.

If reports are transmitted by e-mail, hard copies should be made for enduring record purposes. Those receiving reports should confirm receipt.

Samples and testing

Architects may call for samples of various components and materials used in the building to be submitted for approval, in order that they can satisfy themselves that they meet the specified requirements of the employer and, where applicable, the local authority. Some of the items, of which samples would normally be required, are:

- external cladding materials such as facing bricks, artificial and natural stone, precast concrete, marble, terrazzo, slates and roofing tiles
- internal finishes such as timber, joinery, ironmongery, floor tiles and wall tiles
- services such as plumbing components, sanitary goods and electrical fittings.

In addition to samples of individual components or materials, architects will often require sample panels to be prepared on the site to enable them to judge the effect of the materials in the situations in which they will be used, for example, a panel of facing bricks to demonstrate a particular bond or pattern, the colour of mortar and type of pointing.

It is also quite normal to require laboratory tests of basic materials such as concrete. The testing of concrete should be carried out on a regular basis. Cubes should be cast from each main batch of concrete used, and each cube should be carefully labelled and identified. An approved laboratory should carry out the tests and the test reports should be submitted by the contractor to the architect or structural engineer. Full instructions for such testing procedures are usually included in the specification.

The crushing strength of bricks may also be tested in a laboratory although unless the design requirements are particularly stringent a certificate from the manufacturer giving their characteristics may well suffice.

If required, manufacturers of external cladding will provide test rigs not only to investigate and demonstrate the appearance and structural performance of their products but also to test for water penetration under simulated wind pressure.

British Standard Specifications, Agrément Certificates and Codes of Practice are specified for many building materials, components and processes, and in carrying out tests it is essential to refer to the appropriate standard or code to ensure that the requirements are complied with. In addition, the British Standard Specifications set down acceptable tolerances for manufactured goods. Copies of all relevant standards and codes should always be kept on site by the clerk of works or the contractor.

The architect is empowered under most forms of contract to order tests and also the removal of defective work from site – a notice merely condemning the work is insufficient, and clauses of contracts (for example clause 3.18.4 SBC) are formulated to ensure fair play.

Considerate Constructors Scheme

Constructing a building makes a significant impact on the quality of life in the environment around the site. The impact is particularly significant in city centres, and no more so than in the City of London. In 1987, the Corporation of London pioneered what they called the Considerate Contractors Scheme and in 1997 the Construction Industry Board launched a similar scheme, known as the Considerate Constructors Scheme. This latter scheme is a voluntary code of practice, which seeks to encourage building, demolition and civil engineering contractors to carry out their operations in a safe and considerate manner. The Considerate Constructors Scheme, which is operated by the Construction Confederation, has now been adopted across the country. Annual awards are made under the scheme to recognise and reward commitment on the part of contractors to raise standards of site management, safety and environmental awareness beyond their statutory duties. The Code of Practice for the Considerate Constructors Scheme includes the following criteria:

- **Consideration** All activities are to be carried out with positive consideration for the needs of traders and businesses, site personnel and visitors, and the general public. Special attention is to be given to the needs of those with disabilities.
- **Environment awareness** Noise from construction work, site personnel and other sources is to be kept to a minimum at all times. Local resources should be used wherever possible. Waste management and the avoidance of pollution by recycling surplus materials are to be encouraged.
- **Cleanliness** The working site, safety barriers, lights and signs are to be maintained in a clean and safe condition. Surplus materials and rubbish must not be allowed to accumulate on site or spill over onto adjacent property. Dust from all activities is to be kept to a minimum.
- **Good neighbours** Full and regular consultation with neighbours regarding programming and site activities is to be maintained from pre-start to completion of activities.
- **Respect** All site personnel are to wear respectable and safe dress, appropriate to the weather conditions, and are to be instructed in dealing with the general public. Lewd or derogatory behaviour and language are not to be tolerated. The contractor is to take pride in the management and appearance of the site and surrounding environment.
- **Safety** Movements of site personnel and vehicles are to be carried out with great care and consideration for traders and businesses, site personnel and visitors, and the general public. No building activity is to be a security risk to others.
- **Responsibility** The contractor is to ensure that all persons working on the site understand and implement the Code of Practice.

- **Accountability** Posters are to be displayed clearly around the site, giving names and telephone numbers of staff who can be contacted to respond to issues raised by those affected.

Site safety

Health and Safety Policy

Construction sites should be safe places for those who work on them, and for those who are neighbours. The route to site safety is through the Health and Safety Policy, which falls under CDM94 and the Health and Safety at Work etc. Act 1974 (HASAW).

Section 2(3) of HASAW, states:

> *'Except in such cases as may be prescribed, it shall be the duty of every employer to prepare and as often as may be appropriate revise a written statement of his general policy with respect to the health and safety at work of his employees and the organisation and arrangements for the time being in force for carrying out that policy, and to bring the statement and any revision of it to the notice of all his employees.'*

All employers who employ five or more employees at any one time must prepare such a statement, the main purpose of which is to identify risks involved at work, to identify the precautions and to show who is responsible for carrying out those precautions. The essential ingredients of a good policy statement are:

- a statement of the employer's general policy with regard to health and safety
- details of the organisational/managerial hierarchy for carrying out the policy
- details of the practical arrangements for carrying out the policy
- the name of the person in authority who is responsible for fulfilling the policy.

Although each company is responsible for drafting its own policy statement according to its own needs a model policy document is available from the Health and Safety Executive.

The law only requires safety policy statements to cover the health and safety of employees. However, contractors should state their strategy for protecting other people who could be put at risk by their activities, such as sub-contractors, customers and the public. If the activities of others on site, such as sub-contractors, could put employees at risk, contractors will need to consider how these risks are to be avoided and to cover this aspect in their statement.

If there is a safety committee, its constitution and terms of reference should be clearly established. Further information on safety committees can be found in the Safety Representatives and Safety Committees Regulations 1977 SI 1977 No 500 (as amended) and the Safety Representatives and Safety Committees booklet (ISBN 0 7176 1220 1), which contains the regulations.

Issues to be considered by the contractor when drafting his safety policy statement are many and various.

Generally

- The statement should express a commitment to health and safety and make clear the contractor's obligations towards employees.
- The statement should record which person in authority is responsible for ensuring that it is implemented and regularly reviewed.
- The statement should be signed and dated by a partner or senior director.
- The statement should take account of the views of managers, supervisors, safety representatives and the safety committee.
- The duties set out in the statement should be discussed in advance with the people concerned and accepted by them to ensure that they understand how their performance is to be assessed and what resources they have at their disposal.
- The statement should make it clear that co-operation on the part of all employees is vital to the success of the contractor's health and safety policy.
- The statement should record how employees are to be involved in health and safety matters, for example by being consulted, by taking part in inspections and by sitting on a safety committee.
- The statement should show clearly how the duties for health and safety are allocated and describe the responsibilities at different levels.
- The statement should record who is responsible for the following matters (including deputies where appropriate):
  — reporting investigations and recording accidents
  — fire precautions, fire drill, evacuation procedures
  — first aid
  — safety inspections
  — the training programme
  — ensuring that legal requirements are met, for example regular testing of lifts
  — notifying accidents to the health and safety inspector.

Site safety

- Delivery access, timing and loading restrictions
- Storage compound location
- Inflammable product storage
- De-commissioning, storage or disposal of any existing equipment
- Position of tool store, welfare facilities and shared facilities
- Security and control of the access to high-risk work areas, machinery rooms, control equipment and high voltage switch gear
- Scaffolding and protection arrangements

- Permits-to-work for flame producing or oxy-acetylene cutting gear
- Control of Substances Hazardous to Health (COSHH) assessments of dusty works or substances used in occupied or shared work areas
- Temporary lighting requirements or specification of power locations and usage
- Risk assessments of work operations, relevant in detail and scale to the project in progress
- Method statements relating to the implementation of high risk elements of the work, giving details and descriptions relevant to the size and scale of the works
- Commissioning logs and partial hand-over arrangements
- Lift usage restrictions/authorised user list
- Arrangements for dealing with fire prevention
- Training and competence certificates
- First aid arrangements
- Emergency contact list
- Named management and their respective responsibilities
- Details of how the archived information will relate to the safety file and what operation and maintenance information will be given to the principal contractor during the project
- Keeping the workplace, including staircases, floors, ways in and out, washrooms etc. in a safe and clean condition by cleaning, maintenance and repair.

Plant and substances

- Maintenance of equipment such as tools and ladders in a safe condition
- Maintenance and proper use of safety equipment such as helmets, boots, goggles, respirators, etc.
- Maintenance and proper use of plant, machinery and guards
- Regular testing, maintenance and emergency repair of lifts, hoists, cranes, pressure systems, boilers and other dangerous machinery
- Maintenance of electrical installations and equipment
- Safe storage, handling and, where applicable, packaging, labelling and transport of dangerous substances
- Controls on work involving harmful substances such as lead and asbestos
- The introduction of new plant, equipment or substances into the workplace – by examination, testing and consultation with the workforce.

Other hazards

- The wearing of ear protection, and control of noise at source
- Preventing unnecessary or unauthorised entry into hazardous areas
- The lifting of heavy or awkward loads
- Protecting the safety of site staff against assault when handling or transporting the employer's money or valuables

- Special hazards to site staff when working on unfamiliar sites, including discussion with site manager where necessary
- Control of works transport, such as forklift trucks, by restricting use to properly trained, experienced operatives.

Emergency procedures

- Ensuring that fire exits are marked, unlocked and free from obstruction
- Maintenance and testing of fire-fighting equipment, fire drills and evacuation
- Names and location(s) of persons responsible for first aid and any deputies, and the location of all first aid boxes.

Communication procedures

- Giving site staff information about their general duties under HASAW and specific legal requirements relating to their work
- Giving site staff necessary information about all substances, plant, machinery and equipment with which they come into contact
- Discussing with sub-contractors, before they come on site, how they can plan to do their job with a view to minimising risk, and how to deal with unavoidable hazards that they may create for site staff and that their own employees may encounter on site.

Training procedures

- Ensuring that all employees, supervisors and managers have the necessary training to enable them to work safely and to carry out their health and safety responsibilities efficiently.

Supervision procedures

- Supervising site staff so far as is necessary for their safety – especially young workers, new employees and employees carrying out unfamiliar tasks.

Checking procedures

- Making regular inspections and checks of the workplace, machinery, appliances and working methods.

Fire precautions on site

Fire is a serious and ever-present danger on site and is a key issue when considering site safety. Fires spread rapidly and can endanger life and cause massive damage to property, including collateral damage to neighbouring buildings and equipment. They are a more frequent occurrence than might be thought, there being several construction site fires of some magnitude somewhere every day. Fire precautions depend on the nature and location of the project. Erecting a simple steel-framed building in a rural location will require only simple precautions because fire risks, of occurrence and effect, are minimal. Higher risk work, for example refurbishing floors in an occupied office block, will need many more precautions because the risk of fire occurring, the difficulties of escape and the potential for injury and loss of life are all that much greater (the sites of conservation projects are particularly hazardous). In any event, fire alarm points and local extinguishers (to enable escape, not to fight the fire) must be available throughout the site to comply with regulations.

Regulatory control

The Construction (Health, Safety and Welfare) Regulations 1996 require contractors to take measures both to prevent fires occurring and to ensure that everyone on site, including visitors, is protected if fires happen. From the outset the CDM94 requires all those designing, planning and constructing projects to take account of construction site fire safety in their thinking at all stages.

For larger projects, the Joint Code of Practice on the Protection from Fire of Construction Sites and Buildings Undergoing Renovation (Joint Fire Code) is published by the Construction Confederation and the Fire Protection Association with more detailed and prescriptive requirements.

The Joint Fire Code

This Code applies to works of £1m in value and over. Works of £25m in value and over are termed 'Large Projects' and for these the Code provides additional requirements.

In the sixth edition of the Joint Fire Code, which was published in 2006, it states in clause 2 that:

> *'Non-compliance with the Code by the Construction Industry, by those who procure construction and by construction industry professionals could result in insurance ceasing to be available or being withdrawn resulting in a possible breach of construction contracts which require the provision of such insurance.'*

Failure to comply with the Joint Fire Code might lead to the insurance underwriters withdrawing cover for a project. This would be a great risk to contractor and employer alike. If a project were not covered by insurance, and the contractor were to inadvertently start a fire, the employer might have to begin legal proceedings to recover damages. As most contractors operate using a fluid cash flow, there would be no guarantee that the contractor could sustain an action for damages where reinstatement costs are high. It is likely that the contractor would become insolvent and cease trading.

Clauses 6.14 to 6.16 of the SBC provide for compliance with the Joint Fire Code by both the employer and the contractor when it applies to their particular contract. The SBC also makes provision to ensure that any 'Remedial Measures' required by an insurer to rectify a breach of the Joint Fire Code are carried out timeously.

An example of the code's requirements is the use on site of high-quality flame retardant protection materials approved to Loss Prevention Standard LPS 1207.

Preventing fire

Basic precautions to prevent fires on construction sites are as follows:

- Store LPG cylinders and other flammable materials properly. LPG cylinders should be stored outside buildings in well ventilated and secure areas. Other flammable materials such as solvents and adhesives should be stored in lockable steel containers. Care must be taken to avoid leaks.
- Carefully control hot work such as welding by using formal permit-to-work systems.
- Do not leave tar boilers unattended.
- Keep an orderly site and make sure rubbish is cleared away regularly.
- Avoid unnecessary stockpiling of combustible materials such as polystyrene.
- Take special precautions in areas where flammable atmospheres may develop, for example when using volatile solvents or adhesives in enclosed areas.
- Avoid burning waste materials on site wherever possible. Never use petrol or similar accelerants to start or encourage fires.
- Make sure everyone obeys site rules on smoking.

Means of escape

Construction sites can pose particular problems because the routes in and out may be incomplete and obstructions may be present. Open sites usually offer plentiful means of escape and special arrangements are unlikely to be necessary. In enclosed buildings people can easily become trapped, especially where they are working above or below ground level. In such cases means of escape need careful consideration. The contractor must make sure that:

- Wherever possible, there are at least two escape routes in different directions.
- Travel distances to safety are reduced to a minimum.
- Enclosed escape routes, for example corridors or stairwells, can resist fire and smoke ingress from the surrounding site. (Where fire doors are needed for this, the contractor must ensure they are provided and kept closed; self-closing devices should be fitted to doors on enclosed escape routes.)
- Escape routes and emergency exits are clearly signed.
- Escape routes and exits are kept clear.
- Emergency exits are **never** locked when people are on the site.
- Emergency lighting is installed if necessary to facilitate escape. (This is especially important in enclosed stairways in multi-storey structures, which will be in total darkness if the normal lighting fails during a fire.)
- An assembly point is identified where everyone can gather and be accounted for.

Fire-fighting equipment

The equipment needed depends on the risk of fire occurring and the likely consequences if it does. It can range from a single extinguisher on small low-risk sites to complex fixed installations on large and high-risk sites. In any event the contractor must ensure that:

- Fire-fighting equipment is located where it is really needed and is easily accessible.
- The location of fire-fighting equipment and instructions on its use are clearly indicated.
- The right sorts of extinguishers are provided for the type of fire that could occur. (A combination of water or foam extinguishers for paper and wood fires and CO^2 extinguishers for fires involving electrical equipment is usually appropriate.)
- The equipment provided is maintained and works. (A competent person, normally from the manufacturer, should check fire-fighting equipment regularly.)
- Those carrying out hot work have appropriate fire extinguishers with them and know how to use them.

Emergency plans

The purpose of emergency plans is to ensure that everyone on site reaches safety if there is a fire. Small and low-risk sites only require very simple plans, but higher risk sites will need more careful and detailed consideration. An emergency plan should:

- be available before work starts on site
- be up to date and appropriate for the circumstances concerned
- make clear who does what during a fire
- be incorporated in the construction phase health and safety plan when CDM94 applies
- work if it is ever needed.

Providing information

Fire action notices should be clearly displayed in locations where every site operative can see them. For example, most notices are displayed at site entrances, plant and material distribution points, canteen areas and wash room facilities. Site operatives should be made aware of the locations where information can be found. This is usually done during a site induction meeting prior to an individual commencing work on site.

This list should be amplified or adapted according to the nature of the project and may serve the architect or engineer. Some of the items should be observed jointly with other consultants.

General:

- In all cases check that the work complies with the latest drawings and specification, with the latest requirements of the statutory undertakers and with the building regulations. Ensure that all information is complete.

Preliminary works:

- scaffolding and the Building Act 1984
- location of workmen's canteens, builder's offices, etc.
- suitability and location of clerk of works' or site architect's office
- removal of top soil and location of spoil heaps
- perimeter fencing or hoardings
- protection of rights of way
- protection of trees and other special site features
- protection of materials on site
- party-wall agreements and protection of adjoining property
- site security generally
- agreement on bench mark or level pegs
- agreement on setting out (responsibility of the contractor).

Demolition:

- extent
- adequacy of shoring
- preservation of certain materials and special items (to be listed in the specification or bills of quantities).

Excavation and foundations:

- widths of trenches
- depths of excavations
- nature of ground in relation to trial hole report
- stability of excavations
- pumping arrangements
- risk to adjoining property and general public
- quality of concrete and thickness of beds
- suitability of hardcore (freedom from rubbish)
- suitability of sand and ballast (freedom from loam and correct grading)
- damp-proof membranes and tanking
- ducts, drains or services under building
- size, bending, spacing and placing of reinforcement
- suitability of material for and consolidation of backfilling
- depths of piles and driving/boring conditions.

Drainage:

- depths of inverts and gradients of falls
- thickness and type of bed and jointing of pipes
- quality of bricks for manholes and rendering thereto
- testing of drains and manholes.

Brickwork, blockwork and concrete masonry:

- approval of sample panels of facings and fairface work
- quality and colour of mortar and pointing
- test report on crushing strength where necessary
- BS certificates on load-bearing blocks
- position and type of wall ties

Example 25.1 Standard checklist for site inspections.

- cleanliness of cavities and wall ties in external walls
- setting-out and maintenance of regular vertical and horizontal joints
- type, quality and placing of damp proof courses
- setting-out and fixing of door frames, windows, etc.
- bedding and levels of lintels over openings
- expansion joints.

In-situ concrete:
- setting out and stability of shuttering
- shuttering type to achieve finish specified
- setting out of reinforcement, fixings, holes and water bars
- mix for and procedure of taking test cubes
- curing of concrete and striking of shuttering
- vibration
- use of additives.

Precast concrete:
- size and shape of units
- quality of fit and finish
- position of fixings, holes, etc.
- damage in transit and erection.

Carpentry and joinery:
- freedom from loose knots, shakes, sapwood, insect attack, etc.
- dimensions within permissible tolerances
- application of timber preservatives and primers
- storage, stacking and protection from weather
- jointing, bolting, spiking and notching of carpenter's timber
- spacing of floor and ceiling joists and position of trimmers
- spacing of battens, position of noggings
- weather throatings and cills to doors, windows, etc.
- jointing, machining and finish of manufactured joinery
- fire rating.

Roofing:
- pitch of roof
- spacing of rafters and tile battens
- approval of under felt
- approval of roofing materials and fixings
- pointing to verges and bedding of ridges, etc.
- falls to outlets on flat roofs
- thickness and fixing of insulation under finish
- eaves details and ventilation
- correct formation of flashings, etc.
- ventilation.

Cladding:
- vapour barriers and insulation
- regularity of grounds for sheet materials
- location and quality of fixings
- laps, tolerances, and positions of joints
- setting-out and jointing of mullions and rails
- handling and protection of panel materials
- specification and application of mastic
- flashings, edge trims and weather drips

Example 25.1 (*Cont.*)

- entry and egress of moisture
- prevention of electrolytic action
- location and type of movement joints.

Steelwork:
- sizes of steel
- rivets and welding
- position of members
- plumbing, squaring and levelling of steel frame
- priming and protection.

Metalwork:
- sizing and spacing of members
- galvanising and rustproofing
- stability of supports, including caulking or plugging
- isolation from corrosive materials, etc.

Plumbing and sanitary goods:
- ensuring that sanitary goods are free from cracks and deformities
- location, venting and fixing of stack pipes
- falls to waste branches
- use of traps
- jointing of pipes
- smoke and/or water tests
- location and accessibility of valves, stop-cocks
- drain-down cocks at lowest points
- access to traps and rodding eyes.

Heating, hot water and ventilation installations:
- types of boiler, cylinder, tanks, fans, etc.
- types of pipes
- position and type of stop valves
- position of pipe runs and ventilation ducting
- insulation of pipework and ducting
- identification and labelling of pipes, valves, etc.

Electrical installation:
- types of switchgear, distribution board, motor drives, etc.
- types of switches, socket outlets, fuses, cables, etc.
- location of outlet points
- earthing of installation
- lightning conductor installation
- runs of cables and quality of connections, etc.
- labelling and identification of switchgear, etc.

Specialist installations:
- drawings of specialist installations
- drawings of builder's work in connection with specialist installations
- power and plant to be provided
- access for equipment and provision of adequate working space
- temporary support, loading on structure, lifting tackle, etc.
- attendance on site and sequence of work.

Paving and floor tiling:
- approve materials
- quality of screeds to receive flooring
- junctions of differing floor finishes

Example 25.1 (*Cont.*)

- regularity of finish
- falls to gulleys, etc.
- skirtings and coves
- expansion joints
- types of tile bedding and grouting materials.

Plastering and wall tiling:
- storage of materials
- plaster mixes
- preparation of surface
- true surfaces and arrises
- fixing of plasterboard
- filling and scrimming of joints in plasterboard
- adequate hacking of or bonding plaster on concrete
- regularity of finish
- types of tile bedding and grouting materials.

Suspended ceilings:
- type of suspension and tile
- height of ceiling and setting out
- location and co-ordination of light fittings, sprinkler outlets, ventilation grilles, etc.
- access panels
- finish trim for curtains, blinds, etc.
- fire barriers.

Glazing:
- quality of glass and freedom from defects
- integrity of sealed units
- structural capability
- type and/or thickness
- depth of rebates
- glazing compound, fixing of glazing beads, etc.

Painting and decorating:
- approval of materials
- preparation of surfaces and freedom from damp
- ensuring that partly concealed surfaces are properly finished
- ensuring that finished work is free from runs, brush marks, etc.
- opacity of finish, etc.

Cleaning down and handing over:
- windows cleaned and floors scrubbed
- sanitary goods washed and flushed
- painted surfaces immaculate
- doors correctly fitted, windows not binding or rattling
- ironmongery complete and locks and latches operating correctly
- correct number of and suiting of keys and security cards
- connection of services, provision of meters, etc.
- commissioning of all mechanical engineering plant, balancing of air-conditioning, etc.
- plant maintenance manuals, plant room service diagrams, etc.
- operation of security, communication and fire protection systems
- removal of protective tapes and films etc.
- cleaning or replacement of air filters in the HVAC system
- cleaning of lighting reflectors where uncovered.

Example 25.1 (*Cont.*)

INSTITUTE OF CLERKS OF WORKS

CLERK OF WORKS PROJECT REPORT

NO 6

PROJECT: Shops & offices REF:

ADDRESS: Newbridge Street Borchester

Architect/Contract Administrator Reed & Seymore

Week/~~Month~~ Ending 14 December 2001

Contract Start Date 2 November 2001

Main Contractor Leavesden Barnes

Contract Completion Date 13 September 2002

Clerk of Works Phone/Fax No

Progress +/- to Programme

TRADES	Mon	Tues	Wed	Thur	Fri	Sat	Sun
Site Staff	3	3	2	3	3		
Groundworkers	2	2	2	2	2		
Steelfixers							
Steel Erectors	2	2	2				
Concretors							
Drainlayers	1	1	1	1	1		
Machine Operators							
Carpenters	2	2	2	1	2		
Scaffolders							
Bricklayers	4	4	4	2	2		
Roof Finishers							
Wall Cladding							
Window Fixers							
Glaziers							
Floor Screeders							
Plasterers							
Tilers-Wall/Floor							
Dryliners/Partitions							
Ceiling Fixers							
Decorators							
Floor Finishers							
Heating/Ventilation							
Plumbing							
Electricians							
Hard/Soft Landscape							
Roadworks							
Public Services							
Site agent	1	1		1	1		
TOTAL	15	15	13	10	11		

Contractors Labour Returns May Be Substituted

WEATHER REPORT	AM	°C	PM	°C	Time Lost
Mon	Fine	5	Fine	6	
Tues	"	4	"	5	
Wed	overcast	4	overcast	4	
Thur	heavy rain	4	rain	3	1 day
Fri	Fine	3	Fine	3	
Sat					
Sun					

Stoppages (Hours)

Total to Date

Visitors (Include Statutory Inspectors)	Date
Building Inspector	10 Dec
~~Project architect~~	12 Dec
Health & Safety Inspector	12 Dec

Delays including Defective Work (Action Taken)

3m length of drainage trench collapsed (NE corner) 12 Dec. Re-excavation required with sheet piling. 1 day lost.

Site Directions Issued

No	Item	Date
4	Manhole 3 repositioned	10 Dec

Drawings/Information Received on Site

Roof details - dwg nos. 456/55A, 56B, 57

Drawings/Information Required on Site

Finishes details

Plant/Materials Delivered to Site or Removed

Steelwork delivered

General Comments

Steelwork primer badly scratched

Example 25.2 Clerk of works project report.

	Progress to date	Programme		Progress to date	Programme		Progress to date	Programme
	%	%		%	%		%	%
Preliminaries			Blockwork Internal			Electrical 2nd Fix		
Excavation	80	85	Cladding/Curtain Wall			H & V 2nd Fix		
Shutter/Reinf			Windows-glazing			Ceiling Grid/Tiles		
Concrete Structure			Joinery 1st Fix			Decoration		
Steel Erection	15	15	Plastering			External Works		
Main Drainage m/h	50	60	Drylining/Partitions			Hard/Soft Landscape		
Floor Construction			Floor Screeds			Roadworks		
Floors Suspended			Plumbing 1st Fix			Mains; Gas-Electrical		
Roof Structure			Electrical 1st Fix			Mains; Water-Telecoms		
Roof Coverings			H & V 1st Fix			Lifts		
Drainage fw. sw.	25	30	Wall/Floor Tiles			Alarm/Computer Systems		
Brickwork External			Plumbing 2nd Fix			Defects/Handover		

GENERAL REPORT: Summary of Work Proceeding

Excavations largely completed
Drainage progressing well – but see note under "delays"
Building inspector satisfied with footings
Steelwork started this week

Site Conditions/Cleanliness/Health & Safety: Action Taken

Conditions generally good due to recent fine weather
Reason for collapse of drainage trench being investigated

Enclosures: G.C. Labour Return [ ] Site Directions [✓] Reports (State) [ ]

OFFICE ACTION:

Distribution as Agreed

Client [✓] Architect [✓] Project Manager [·] Quantity Surveyor [✓] Office [ ] Others [ ]

Clerk of Works Harry Hemmings Date 17 December 2001

Example 25.2 (*Cont.*)

Chapter 26
Instructions

Provided that the procedures set out in the preceding chapters are followed, it should be possible to have available complete sets of drawings, specification notes and quotations from sub-contractors and suppliers when the tender documents are prepared. This in turn will mean that, as soon as the contract is placed, contractors can be handed all the necessary information to enable them to build the project.

Architect/contract administrator's instructions

However, the ideal circumstances outlined above are not the norm, and even when fully finalised information is available for incorporation into the contract documents, it will still be necessary, from time to time, for the architect/contract administrator to issue further drawings, details and instructions. These are collectively known as architect's instructions. The SBC conditions of contract set out those matters in connection with which the architect is empowered to issue such instructions and these are:

Clause	Description
2.10	Levels and setting out of the Works
2.12	Provision of further drawings or details
2.15	Discrepancies in or divergences between documents
2.17	Divergences from Statutory Requirements
2.18	Emergency compliance with Statutory Requirements
2.38	Schedules of defects and instructions
3.14	Instructions requiring a variation
3.15	Architect's instructions – postponement of work
3.16	Instructions on provisional sums
3.17	Inspection – tests
3.18	Work not in accordance with the Contract
3.19	Workmanship not in accordance with the Contract
3.21	Exclusion of persons from the Works

3.23	Instructions on antiquities
4.5.2.1	Contract Sum and Adjustments – Final adjustment
5.3	Schedule 2 Quotation
6.5.1	Contractor's insurance of liability of Employer
6.15	Breach of Joint Fire Code – Remedial Measures.

The procedure for the issue of architect/contract administrator's instructions is set out in SBC clause 3.12 and can be summarised as follows:

- Any instruction issued by the architect/contract administrator must be in writing, but
- should the architect/contract administrator issue an instruction other than in writing it is of no effect unless, within seven days of such an instruction being given:
  - — it is confirmed in writing by the architect/contract administrator, at which time it becomes effective, or
  - — it is confirmed in writing by the contractor and is not dissented from by the architect/contract administrator within a further period of seven days, at which time it becomes effective; however
- if neither the architect/contract administrator nor the contractor confirms an instruction issued other than in writing but the contractor complies with the

Any instruction issued by the architect must be in writing

instruction, the architect/contract administrator can, at any time up to the issue of the final certificate, confirm that instruction in writing, at which time the instruction becomes effective.

Whilst it is worth noting that SBC clause 3.13 allows the contractor to question the contractual validity of any architect/contract administrator's instruction and requires the architect/contract administrator to answer 'forthwith', it should also be noted that, if the contractor does not comply with a valid architect/contract administrator's instruction, clause 3.11 allows the employer to employ others to give effect to the instruction and recover all costs so incurred from the defaulting contractor.

If a clerk of works is employed on the project and issues any directions to the contractor, such directions are effective only if they are issued in regard to a matter in respect of which the architect/contract administrator is empowered to issue instructions, and also if they are confirmed in writing by the architect/contract administrator within two working days (not, it will be noted, within seven days as is provided for the confirmation by architects/contract administrators of their own instructions issued other than in writing). Standard forms are available for such directions from the clerk of works.

It is good practice for all instructions from the architect/contract administrator to the contractor to be issued or confirmed on standard forms. An example of such a form, which is published by RIBA Enterprises Ltd, is given as Example 26.1 at the end of this chapter.

It is essential that instructions should be clear and precise and, where revised drawings are issued, the date and reference of the particular revision should be specifically referred to. Instructions emanating from other members of the design team must not be given directly to the contractor but must be issued via the architect/contract administrator as architect/contract administrator's instructions. Copies of architect's instructions should be distributed to all of the following:

- contractor
- employer or project manager or employer's representative
- planning supervisor
- quantity surveyor
- other members of the design team
- clerk of works
- any sub-contractor affected by the instruction.

The SBC also makes provision for the architect/contract administrator to issue instructions to the quantity surveyor and these are as follows:

Clause	**Description**
4.18.2	To prepare a statement specifying the details of the retention deducted in arriving at the amount stated as due in interim certificates.
4.23 & 3.24	To ascertain the amount of loss and/or expense incurred by the contractor.

Architect's Instruction

Issued by: Ivor Barch Associates
address: Prospects Drive, Fairbridge

Employer: Cosmeston Preparatory School
address: Fairbridge

Contractor: L&M Construction Ltd
address: Ferry Road, Fairbridge

Works: New School Library
situated at: Park Street, Fairbridge

Contract dated: 14 December 1998

Job reference: IBA/94/20

Instruction no: 4

Issue date: 18 February 1999

Sheet: 1 of 2

Under the terms of the above-mentioned Contract, I/we issue the following instructions:

		Office use: Approximate costs £ omit	£ add
1.	INCOMING GAS MAIN Accept the quotation ref. no. 8438/63 dated 6 January 1999 received from EuroGas in the sum of £248.00 for the new incoming gas main and supply and installation of gas meter. A copy of their quotation is attached. This sum is to be set against the provisional sum of £400 included under ref. 3/2G in the bill of quantities.	400.00	248.00
2.	(a) HIP TILES OMIT: Farland concrete third round hip tiles, bill of quantities ref. 5/15E. ADD in lieu: Red Bank 300mm long red Terracotta third round segmental ridge tiles, list no. 259.	372.00	514.00
	(b) HIP IRONS OMIT: 4 no. hip irons, bill of quantities ref. 5/15F. ADD in lieu: 4 no. Red Bank 300mm long red Terracotta Scroll hip finial tiles 225mm diameter. Hip tiles and finial tiles to be obtained from Farbridge Bank Manufacturing Co. Ltd.	–	–
3.	PICTURE-HANGING FACILITIES IN ACTIVITIES ROOM Supply and fix aluminium picture rail approx. 12m long and 100 no. type (b) U-shaped hooks obtainable from Library Aids of Thawbridge. Picture rail to be fixed in location shown on attached drawing no. IBA/94/20/61 at 2050mm from finished floor level. (continued)	–	80.00
	sub-total:	772.00	842.00

To be signed by or for the issuer named above — Signed *Ivor Barch*

Amount of Contract Sum £
± Approximate value of previous Instructions £
Sub-total £
± Approximate value of this Instruction £
Approximate adjusted total £

Distribution
☐ Contractor ☐ Quantity Surveyor ☐ Clerk of Works ☐
☐ Employer ☐ Structural Engineer ☐ Planning Supervisor ☐
☐ Nominated Sub-Contractors ☐ M&E Consultant ☐ ☐ File

F809 for JCT 98 / IFC 98 / MW 98

Example 26.1 Architect's instruction.

Issued by: Ivor Barch Associates
address: Prospects Drive, Fairbridge

Instruction
continuation

Job reference: IBA/94/20

Instruction no: 4

Issue date: 18 February 1999

Sheet: 2 of 2

	£ omit	£ add
Brought forward:	772.00	842.30
4. VELUX ROOFLIGHTS OMIT: single glazing with 6mm thick clear polycarbonate sheet, bill of quantities ref. 5/25B. ADD in lieu: single glazed in 10mm thick Clerestory Glass Ltd. 'Kevlarplate' float clear.	–	–
5. MORTAR MIXES This is to confirm Clerk of Works Direction no. 6 that mortar mixes are as follows: Blockwork below dpc 1:3 Blockwork above dpc 1:1:6 Brickwork below dpc (below ground level) 1:3 Brickwork below dpc (above ground level) 1:3 (pointing in 2:1:9) Brickwork above dpc 2:1:9 Coloured mortar is to be Zilcon mortar, ref. Y73.	–	–
6. REDUCED LEVELS This is to confirm Clerk of Works Direction no. 6. Excavate to revised levels to remove unsuitable material. The 'as dug' levels as shown on the record made by the site foreman and clerk of works are agreed.	–	–
This instruction is issued in accordance with clauses 13.3, 13.2 and 4.1 of the contract. (Items nos. 4, 5 and 6 are to be at no cost to the Employer.)		
To be signed by or for the issuer named above — Signed *Ivor Barch*	772.00	842.30

Amount of Contract Sum	£ 505,096.00
± Approximate value of previous Instructions	£ 500.00
Sub-total	£ 505,596.00
± Approximate value of this Instruction	£ 70.00
Approximate adjusted total	£ 505,666.00

F820 for JCT 98 / IFC 98 / MW 98

Example 26.1 (*Cont.*)

Chapter 27

Variations and Post-Contract Cost Control

Variations

It is important not to confuse architect/contract administrator's instructions with variations. Whilst all variations arise as a consequence of architect/contract administrator's instructions, not all architect/contract administrator's instructions give rise to variations. In the SBC, clause 5 provides the definition of a variation as follows:

5.1 *The term 'Variation' means:*

5.1.1 *the alteration or modification of the design, quality or quantity of the Works, including:*

5.1.1.1 *the addition, omission or substitution of any work,*

5.1.1.2 *the alteration of the kind or standard of any of the materials or goods to be used in the Works;*

5.1.1.3 *the removal from the site of any work executed or materials or goods brought thereon by the Contractor for the purpose of the Works other than work, materials or goods which are not in accordance with the Contract;*

5.1.2 *the imposition by the Employer of any obligations or restrictions in regard to the matters set out in this clause 5.1.2 or the addition to or alteration or omission of any such obligations or restrictions so imposed or imposed by the Employer in the Contract Bills or in the Employer's Requirements in regard to:*

5.1.2.1 *access to the site or use of any specific parts of the site;*

5.1.2.2 *limitations of working space;*

5.1.2.3 *limitations of working hours; or*

5.1.2.4 *the execution or completion of the work in any specific order.*

The matters set out in SBC clause 5.1.2 relate to the working conditions imposed by the employer, particulars of which will have been given in the bills of quantities in accordance with Section A of SMM7R, or in the specification. It should be noted that whilst there is a general obligation, under SBC clause 3.10, for the contractor to comply forthwith with all architect/contract administrator's instructions, clause

3.10.1 allows that the contractor need not comply with an architect's instruction requiring a variation within the meaning of clause 5.1.2 so long as he has given the architect/contract administrator reasonable, written objection to such compliance. The reason for giving the contractor this important right of objection may well be in recognition of the possibility that such a variation could so fundamentally affect the basis on which the contractor tendered that the valuation of variations and direct loss and/or expense provisions within the SBC would not provide adequate recompense.

Valuing variations

All variations, other than those for which a contractor's Schedule 2 Quotation has been accepted under the terms of SBC clause 5.3, fall to be valued by the quantity surveyor in accordance with the Valuation Rules of SBC clause 5.6. These clauses prescribe the five basic methods for the valuation of all work executed by the contractor as the result of architect/contract administrator's instructions as to the expenditure of provisional sums and the execution of work for which an approximate quantity is included in the contract bills and variations.

The method of valuation to be adopted by the quantity surveyor depends upon the nature of the work and the conditions under which it has to be executed. Each basic method of valuation and the nature of the work to which it is to be applied are summarised in Table 27.1.

It must be emphasised that, under SBC clause 5.10, no allowance is to be made in any valuation carried out under SBC clause 5.2 for any effect upon the regular progress of the works or for any other direct loss and/or expense for which the contractor would be reimbursed by payment under any other provision of SBC, for example, under clause 4.23 on loss and expense – matters materially affecting regular progress.

It is interesting to note that when the contract rules for valuing variations are brought into play the quantity surveyor has a unilateral responsibility – the contractor is not involved, apart from being entitled to be present when any measurements are made. Should the contractor not be satisfied with the quantity surveyor's valuation, the only formal recourse is to use the dispute resolution procedures set out in the contract. Practically, of course, the quantity surveyor usually works closely with the contractor's surveyor so that (more often than not) disputes are avoided and an agreed final account is produced.

Dayworks

Works valued at prime cost on a daywork basis effectively involve a cost-plus method of reimbursement and, from a contractor's viewpoint, this is therefore an attractive means of securing payment for variations. For work to be valued at prime

Table 27.1 The five basic methods for the valuation of all variations.

Description of work	Method of valuation
1. Additional or substituted work that can properly be valued by measurement and where the work is of similar character to, is executed under similar conditions as, and does not significantly change the quantity of work set out in the contract bills. Work for which an approximate quantity is included in the contract bills and that quantity is reasonably accurate. Omission of work set out in the contract bills.	Measure in accordance with SMM7R. Value at the rates and prices for the work set out in the contract bills. Allow for any percentage or lump sum adjustments in the contract bills. Allow for any addition to or reduction of preliminary items of the type referred to in SMM7R section A except in the case of work resulting from an instruction as to the expenditure of a provisional sum in the contract bills for defined work.
2. Additional or substituted work that can properly be valued by measurement and where the work is of similar character to work set out in the contract bills but is not executed under similar conditions thereto and/or significantly changes the quantity thereof. Work for which an approximate quantity is included in the contract bills and that quantity is not reasonably accurate. Any work executed under changed conditions as a result of a variation, compliance with an instruction as to the expenditure of a provisional sum or the execution of work for which an approximate quantity is included in the contract bills.	As 1 above plus a fair allowance for the differences arising from the work not being executed under similar conditions to and/or because the variation changes the quantity of the work of similar character in the contract bills.
3. Additional or substituted work that can properly be valued by measurement and where the work is not of similar character to work set out in the contract bills.	Measure in accordance with SMM7R. Value at fair rates and prices. Allow for any percentage or lump sum adjustments in the contract bills. Allow for any addition to or reduction of preliminary items of the type referred to in SMM7R section A except in the case of work resulting from an instruction as to the expenditure of a provisional sum in the contract bills for defined work.
4. Additional or substituted work that cannot properly be valued by measurement.	Provided that vouchers specifying the time spent upon the work, the workmen's names, the plant and the materials employed are delivered to the architect for verification not later than the end of the week following that in which the work is executed, value at the prime cost of such work calculated in accordance with the 'Definition of Prime Cost of Daywork carried out under a Building Contract' issued by the Royal Institution of Chartered Surveyors and the Construction Confederation or the Electrical Contractors' Association or the Heating and Ventilating Contractors' Association, as appropriate, and current at the base date together with percentage additions to each section of the prime cost at the rates set out in the contract bills.
5. Work other than additional, substituted or omitted work. Work or liabilities directly associated with a variation that cannot reasonably be valued by any other method.	A fair valuation.

... the work is of similar character ...

cost on a daywork basis the record sheets must be submitted to the architect/contract administrator or his authorised representative, usually the clerk of works, no later than one week following the week in which the work was carried out. In addition to being serially numbered it is essential that the following information be recorded on each sheet:

- the reference of the architect/contract administrator's instruction authorising the work
- the date(s) when the work was carried out
- the daily time spent on the work, set against each operative's name
- the plant used
- the materials used.

The architect/contract administrator or his authorised representative should, as soon as possible after receiving the sheets, check the accuracy of the records and, if found to be correct, countersign them and pass them to the quantity surveyor who will determine whether or not the recorded work is to be valued at prime cost. It is worth noting that the architect/contract administrator's signature on a daywork sheet does not constitute a variation, nor does it commit the quantity surveyor to having to value the work at prime cost.

It is helpful if the contractor gives the architect/contract administrator advance warning of his intention of recording anything as daywork so that the architect/contract administrator, or the clerk of works on his behalf, can arrange for particular notice to be taken of the resources used.

Cost control

Cost control may be defined as the controlling measures necessary to ensure that the authorised maximum cost of the project is not exceeded. It is a continuous process and follows on from the pre-contract cost control activities discussed in Chapter 16.

It is essential for the design team to establish at the outset the parameters within which the employer requires the construction costs to be controlled. Usually employers have a limit on the amount of money available for expenditure on a project and will insist that this is not exceeded. It has to be said, however, that occasionally the need to achieve a high-quality end product is more important to the employer than cost.

Initially, the authorised expenditure limit for a project will more often than not be the contract sum. However, it is not uncommon for the limit to be varied during construction – for example, an employer, building speculatively, may find that a prospective tenant is prepared to pay more for an enhanced specification, in which case the employer will require the design team to assess the cost and time implications of incorporating the enhanced specification into the project and, if economically viable, the employer will approve an increase in the previously authorised expenditure limit.

As most construction projects are complex it is normal for a contingency amount to be included in the contract sum to cater for expenditure on unforeseen items of work that become necessary during the construction process. The amount of the contingency sum will vary according to the nature of the project and the perceived risk of expenditure on unforeseen items. A refurbishment project, for example, would most likely require a larger contingency provision than would a similar value new building project on a green-field site. It would be normal for the contingency sum to equal between 3% and 5% of the contract sum, although it could be more on a particularly risky project. The primary purpose of the contingency sum is to fund additional work that could not reasonably have been foreseen at design stage, for example additional work below ground; it is not there to fund design alterations, except with the prior approval of the employer. In circumstances where no unforeseen items of work arise, the contingency sum should remain unspent at practical completion but, in practice, this is a rare (indeed almost unheard of) occurrence.

If construction costs are to be controlled successfully, it is essential for there to be good communication between all members of the building team. As it falls to quantity surveyors to maintain construction cost records and provide employers with regular financial reviews, it is essential that they attend all meetings when cost matters are discussed and that they receive copies of all documents that may have some bearing on the cost of any project.

It is also essential that the cost effect of any proposed variations, whether emanating from the architect/contract administrator or from any of the other design consultants, is determined as soon as possible so that steps can be taken to minimise its impact on the contract sum, maintaining overall expenditure within authorised limits before any architect/contract administrator's instruction authorising it is issued. Traditionally this has been done, with varying degrees of accuracy, by the quantity surveyor estimating the cost effect of all proposed variations. However, a contractor's Schedule 2 Quotation means that it is possible to obtain a fixed-price quotation from the contractor for a proposed variation prior to authorising its execution. Whilst using this procedure may take a little longer than the quantity surveyor's estimate, it does provide fixed costs, which should lead to more accurate cost control.

By looking ahead and making early decisions on such matters as the listing of sub-contractors and suppliers and the expenditure of provisional sums, the architect/contract administrator can greatly assist the cost control process.

The regular financial reviews referred to above will normally be produced by the quantity surveyor and sent to the employer and other members of the design team at monthly intervals, often coinciding with the dates of issue of interim certificates. These reviews should show a forecast of the employer's total financial commitment to the contractor, including the reimbursement of direct loss and/or expense but normally excluding VAT, and should, in effect, comprise an estimate of the final adjustment of the contract sum. A list of all the adjustments to be made to the contract sum is set down in SBC clause 4.5 and this can act as a most useful *aide memoire* when preparing a financial review.

A typical financial review is given as Example 27.1

FUSSEDON KNOWLES & PARTNERS
Chartered Quantity Surveyors
Upper Market Street
Borchester BC2 1HH

Financial Review No: 9
Date of Issue: 12 March 2002
Reference: 1234

Works: Shops & Offices, Newbridge Street, Borchester

Contractor: Leavesden Barnes & Co Ltd

Employer: Aqua Products Ltd

Contract Sum: £675,332.00

Approved Expenditure: £680,000

Date of Possession: 11 June 2001

Date for Completion: 26 July 2002

ESTIMATED CURRENT FINANCIAL COMMITMENT MORE THAN APPROVED EXPENDITURE

	£	£
Contract sum	675,332	
Deduct contingencies	20,000	
		655,332
Add estimated value of variations		23,980
		679,312
Add/deduct estimated fluctuations in cost of labour and materials		NIL
		679,312
Add estimated reimbursement of direct loss and/or expense		2,000
Total estimated final cost		681,312
Deduct approved expenditure		680,000
OVERSPENT BY	**£**	**1,312**

ESTIMATED CURRENT FINANCIAL COMMITMENT LESS THAN APPROVED EXPENDITURE

	£	£
Approved expenditure		
Contract sum		
Deduct contingencies		
Add/deduct estimated value of variations		
Add/deduct estimated fluctuations in cost of labour and materials		
Add estimated reimbursement of direct loss and/or expense		
Total estimated final cost		
UNDERSPENT BY	**£**	

Notes: 1. All amounts given above are EXCLUSIVE of fees and Value Added Tax

Signature:

Date: 12 March 2002

Example 27.1 Financial review.

Chapter 28
Interim Payments

Introduction

Prior to 1998 it was a basic principle of English contract law that when a contractor undertook to do work for a fixed sum, he was not due any payment until the whole of the work had been completed, but once the work was complete, he was due full payment of the fixed sum. Whilst this arrangement may have been appropriate for small projects, construction work can involve large sums of money being expended over a number of months or even years. Thus, in order to provide cash flow to the contractor, this basic principle was usually varied by agreement between the parties to allow interim payments to be made as the work proceeded.

The Housing Grants, Construction and Regeneration Act 1996 (the Construction Act) and The Scheme for Construction Contracts (England and Wales) Regulations 1998 (The Scheme) came into effect on 1 May 1998 and changed contract law in respect of what the Construction Act defined as construction contracts, except for those with a residential occupier. Amongst other things, the Construction Act requires that in respect of payments:

- A party to a construction contract is entitled to payment by instalments, stage payments or other periodic payments for any work under the contract unless the duration of the work is less than 45 days. The parties are free to agree the amounts of the payments and the intervals at which, or the circumstances in which, they become due. In the absence of agreement such amounts, intervals or circumstances have to be determined by reference to Part II of The Scheme.

- Every construction contract shall provide a mechanism for determining what payments become due under the contract, and when, and shall also provide for a final date for payment of any sum which becomes due. The parties are free to agree how long the period is to be between the date on which the sum becomes due and the final date for payment. In the absence of such provisions these matters have to be determined by reference to Part II of The Scheme.

- Every construction contract shall provide for the giving of notice by a party, not later than five days after the date on which a payment becomes due from him, specifying the amount of the payment to be made and the basis on which that amount is calculated. In the absence of such provisions, the notice specifying the amount of payment has to be that set down in Part II of The Scheme.
- A party to a construction contract may not withhold payment after the final date for payment unless he has given an effective notice of intention to withhold payment not later than the prescribed period before the final date for payment. The parties are free to agree the length of the prescribed period before the final date for payment. In the absence of such agreement, the period has to be that set down in Part II of The Scheme.
- Where the sum due under a construction contract is not paid in full by the final date for payment and no effective notice to withhold payment has been given, the person to whom the sum is due has the right to suspend performance of his obligations under the contract.
- A provision making payment under a construction contract conditional on the payer receiving payment from a third party is ineffective, unless that third party is insolvent.

All standard forms of contract now incorporate such payment provisions as are necessary to comply with the Construction Act. They also provide for money to be held in trust, as retention both during and at the end of the period of construction, primarily to provide a fund from which the employer is able to recover the cost of having defects repaired which the contractor may be unwilling to put right.

When considering the subject of interim payments it is worth bearing in mind that the contractor has similar obligations to his suppliers and sub-contractors as does the employer to the contractor. Provided that the minimum requirements of the Construction Act are met, interim payment provisions can be in any one of the following forms:

- payments of pre-determined amounts at regular intervals
- payments of pre-determined amounts when the work reaches pre-determined stages of completion
- payments of amounts at regular intervals calculated by a detailed valuation process as the work proceeds.

Payments of pre-determined amounts at regular intervals

This is a method where, for example, a construction project of £1,200,000 is contracted to take 12 months and it is determined that the contractor will be paid

£100,000 per month, adjusted to account for authorised variations, fluctuations, loss and/or expense and retention. The method has the advantage to both parties of minimal administration and cash flow certainty. However, it can lead to problems if the contractor falls behind programme or executes defective work, whereupon over-payment to the contractor for the work done is a probability, which would certainly lead to problems in the event of the contractor going into liquidation before completing the project.

Pre-determined payments at pre-determined stages

This is a method whereby pre-determined amounts, adjusted as above, become due for payment upon the proper completion of a pre-determined stage of work. For example, the pre-determined amount of, say, £72,000 (duly adjusted) becomes payable upon the proper completion of all work up to damp-proof course level.

Determining the amounts of interim payments in this way keeps down administration costs and, unlike the previous method, does not lead to overpayment to the contractor if he falls behind programme or if he executes defective work. It can, however, be disadvantageous to the contractor as it does not take into account the value of unfixed materials and goods either on or off site. Also the non-completion of a minor, non-critical and low-value element of work in a high-value stage would prevent payment becoming due for the whole stage.

It is interesting to note that the JCT's Design and Build Contract provides for the optional use of this method of payment under clause 4.7 Alternative A – Stage Payments.

Regular payments by detailed valuation

This method of determining the amounts of interim payments is the fairest and most commonly adopted in construction contracts. The implications of defective work, delays, authorised variations, materials on and off site, fluctuations, loss and/or expense and retention can all be readily taken into account in each payment. This procedure does, however, require a regular and substantial input from most members of the building team.

The SBC adopts this method of payment and sets out the procedures to be followed in considerable detail in Section 4 – Payment.

Certificates and payments under the SBC

The obligations of the employer's consultants and the parties to the contract, as set out in Section 4, are briefly as follows.

The architect/contract administrator

The architect/contract administrator has to issue interim certificates stating the amount due to the contractor, to what that amount relates and the basis of calculation of that amount:

- on the dates provided for in the contract particulars up to the date of practical completion or to within one month thereafter. (These will not be at intervals of less than one calendar month.)
- after practical completion as and when further amounts are ascertained as being payable to the contractor
- after expiration of the rectification period or upon issue of the certificate of making good, whichever is the later.

Interim certificates are vital to the smooth running of a project and the architect/contract administrator should bear in mind that:

- certificates must be issued at the appointed time and in accordance with the requirements of the contract
- he is legally responsible for the accuracy of his certificates
- he must remain independent in the issuing of certificates so as to be fair to both parties.

The quantity surveyor

If contractors submit an application setting out their gross valuation to the quantity surveyor, the quantity surveyor has to make an interim valuation. If the quantity surveyor disagrees with the gross valuation of the contractors, he has to submit a statement to the contractor at the time of making his valuation and in similar detail to the contractor's application, identifying his disagreement. If clause 4.21 – choice of fluctuation provisions – is stated in the contract particulars as applying, the quantity surveyor has to make an interim valuation before the issue of each interim certificate. Otherwise it is at the architect/contract administrator's discretion whether or not interim valuations are to be made by the quantity surveyor.

The employer

Not later than five days after the date of issue of an interim certificate the employer ought to give written notice to the contractor (in respect of the amount certified due) specifying the amount of the payment proposed, to what that amount relates, and the basis of calculation of that amount. Not later than five days before the final date for

... the architect is legally responsible for the accuracy of his certificates ...

payment of the amount certified due, the employer may give written notice to the contractor specifying the amount proposed to be withheld or deducted from the due amount, the ground(s) for so doing, and the amount attributable to each ground.

If the employer does not give these notices and fails to pay, in full, the amount stated as due in an interim certificate by the final date for payment, that is within 14 days from the date of issue of the interim certificate, then:

- contractors are entitled to be paid simple interest on the overdue amount at 5% per annum above the official dealing rate of the Bank of England
- contractors are empowered, subject to giving the notice required by clause 4.14, to suspend the performance of their obligations under the contract until payment in full occurs
- contractors are empowered, subject to giving the notices required by clause 8.9, to terminate their employment under the contract and they may also start proceedings in the courts for the recovery of the debt.

Whilst employers have the right, in the circumstances prescribed by and subject to giving the notices required by clause 2.31, to deduct liquidated damages from the amount stated as due in any interim certificate, they do not have the right to interfere with the issue of any architect/contract administrator's certificate. If they do so it is a matter for which the contractor may determine his employment under the contract. However, contractors would achieve nothing by determining if it were the final certificate that was obstructed or interfered with, and it is doubtful whether they would achieve much by determining close to or after practical completion. In such circumstances they would, instead, be better off invoking the procedures for the settlement of disputes or differences prescribed by mediation, adjudication, arbitration or legal proceedings.

When it is recorded in the contract particulars that the employer is a 'contractor', the employer must, when making payment, comply with the provisions of clause 4.7 and must do so in accordance with the Construction Industry Scheme (CIS). Certain non-construction businesses or other concerns are required to act as contractors. These include government departments, some public bodies and businesses that have an average annual expenditure in excess of £1 million on construction operations over three years ending with their last accounting date. These organisations are required by HM Revenue and Customs to operate the Construction Industry Scheme.

The contractor

Whilst clause 4.10, subject to any agreement between the parties, allows contractors to submit to the quantity surveyor an application setting out their gross valuation pursuant to clause 4.16, they are under no contractual obligation to assist in any way in the preparation of interim valuations or certificates. It is the architect/contract administrator's responsibility to issue the interim certificates at the correct times and it is the employer's responsibility to make payment to the contractors within the 14 day period.

Not later than five days after the date that payment becomes due the contractor ought to give written notice to each sub-contractor specifying the amount of the payment in respect of their sub-contract works, to what that amount relates, and the basis of calculation of that amount. Not later than five days before the final date for interim payment, the contractor may give written notice to each sub-contractor specifying the amount proposed to be withheld or deducted from the amount due to them, the ground(s) for so doing, and the amount attributable to each ground. If the contractor does not give these notices and fails to pay, in full, the amount stated as due in an interim payment by the final date for payment, that is within 21 days after the date on which they become due, then:

- the sub-contractor is entitled to be paid simple interest on the overdue amount at 5% per annum above the official dealing rate of the Bank of England

- the sub-contractor is empowered, subject to giving the notice required by clause 4.11 of Standard Building Sub-Contract (SBCSub), to suspend the performance of his obligations under the contract until payment in full occurs.

Interim certificates under the SBC

Whether or not there is a contractual requirement for the quantity surveyor to prepare interim valuations for the purpose of ascertaining the amount to be stated as due in an interim certificate, it is normal practice on most projects for this to be done and for the contractor to co-operate.

According to clause 4.10 the amount to be stated as due in an interim certificate is the gross valuation up to and including a date not more than seven days before the date of the interim certificate less:

- any amount which may be deducted by the employer as the retention
- the total amount of any advance payment due for reimbursement to the employer
- the total amount stated as due in previous interim certificates.

Clause 4.16 provides that the gross valuation shall be:

- the total of all the amounts to be included which are subject to retention
- plus the total of all the amounts to be included which are not subject to retention
- less the total of all the amounts to be deducted which are not subject to retention.

The amounts to be included which are subject to retention are:

- the total value of the work, properly executed by the contractor, including variations and, where applicable, any adjustment of that value under fluctuation Option C, but excluding any restoration, replacement, replacement or repair of loss or damage and removal and disposal of debris treated as a variation following a claim on the insurance policy covering the works
- the total value of materials and goods delivered to, or adjacent to the works and due to be incorporated, provided they are reasonably and not prematurely delivered – and have been protected from the weather
- the total value of any listed items, before their delivery to or adjacent to the works (see *Unfixed materials and goods off site* below).

The amounts to be included which are not subject to retention are:

- any amount to be included in interim certificates in respect of
  - — statutory fees and charges
  - — opening up work for inspection and testing

  - patent rights arising out of compliance with an architect's instruction
  - various insurances under clauses 2.6.2, 6.5.3, A.5.1, B.2.1.2 and C.3.1
- any amount due to the contractor by way of reimbursement of loss and/or expense arising from matters materially affecting the regular progress or from the discovery of antiquities
- any amount due to the contractor in respect of any restoration, replacement or repair of loss or damage and disposal of debris which are treated as a variation under paragraphs B.3.5 and C.4.5.2 of Schedule 3
- any amount due to the contractor under fluctuation Options A or B.

The amounts to be deducted which are not subject to retention are:

- any amount in respect of errors in setting out which the architect/contract administrator instructs are not to be amended
- any amount in respect of any defects, shrinkages or other faults which the architect/contract administrator instructs are not to be made good
- any amount in respect of work completed by another person following non-compliance with an architect/contract administrator's instruction in accordance with clause 3.11
- any amount in respect of work not in accordance with the contract which the architect/contract administrator may allow to remain
- any amount allowable by the contractor to the employer under fluctuations Options A or B, if applicable.

Unfixed materials and goods on site

The value of all materials and goods stored on site must be included in the valuation, provided that they are adequately protected and have not been brought to site prematurely. As an extreme example, unless there was some prior agreement on the matter, the architect/contract administrator is entitled to withhold payment for items of furniture brought onto site whilst only work on the foundations is in progress. Any materials or goods included in the amount stated as being due in an interim certificate that has been paid by the employer become the property of the employer and, although the contractor remains responsible for their loss or damage, they must not be removed from the site.

Unfixed materials and goods off site

Only the value of materials or goods stored off site that have been listed by the employer in a list supplied to the contractor and annexed to the contract documents can be considered for inclusion in a gross valuation, and then only if the following criteria are met:

- the contractor has provided the architect/contract administrator with reasonable proof that the property in the listed items is vested in the contractor
- if so stated in the contract particulars, the contractor has provided a bond in favour of the employer from a surety approved by the employer in respect of payment for the listed items
- the listed items are in accordance with the contract
- the listed items are set apart or are clearly and visibly marked, and identify the employer and to whose order they are held and their destination as the works
- the contractor has provided the employer with reasonable proof that the listed items are insured against loss or damage for their full value.

Any listed items included in the amount stated as being due in an interim certificate that has been paid by the employer become the property of the employer and, whilst the contractor remains responsible for their loss or damage, and the cost of their storage, handling and insurance, they must not be removed from the premises where they are stored except for the purpose of their delivery to site.

Retention under the SBC

The contract provides that, in the calculation of the amount to be stated as due in an interim certificate, the employer may deduct 'the retention' from the gross valuation. The total amount of the retention at any one time is ascertained by application of the rules given in clause 4.20 and comprises:

- the total value of work (or sections) properly executed which has not reached practical completion and of materials and goods included in the gross valuation – at the retention percentage
- the total value of work (or sections) which has reached practical completion or has been taken into the employer's possession by agreement with the contractor, but for which a certificate of making good defects has not been issued – at half the retention percentage.

The retention percentage is 3% unless a lower rate is specified in the appendix to the contract. It is to be noted that no retention is held against the value of work for which a certificate of making good defects has been issued.

The treatment of the retention is subject to the rules given in clause 4.18 which provide as follows:

- The employer's interest in the retention is fiduciary as trustee for the contractor, but without an obligation to invest. In other words it is money held by the employer in trust for the contractor.

- The architect/contract administrator has to prepare, or instruct the quantity surveyor to prepare, a statement of the contractor's retention at the date of each interim certificate and copies are to be issued to both the employer and the contractor.
- The employer shall, at the request of the contractor, place the retention in a separate, appropriately designated banking account and certify to the architect/contract administrator, with a copy to the contractor, that this has been done. (This rule does not apply when a local authority is the employer.)
- When employers exercise their right to withhold and/or deduct monies due to them under the terms of the contract against any retention, they must inform the contractor of the amount withheld and/or deducted.

Payments to sub-contractors under the SBC

The SBC contains very few references to a contractor's obligation to pay sub-contractors, which is obviously dealt with within each respective sub-contract. However, under clause 3.9.2, it is stated that each the sub-contract must provide that if the contractor fails to pay properly any amount due to the sub-contractor by the final date for payment, the contractor shall pay interest on the amount not properly paid. The interest that the sub-contractor is entitled to receive is simple interest at the rate of 5% above the official dealing rate of the Bank of England.

Value added tax

Clause 4.6 makes it quite clear that the contract sum is exclusive of value added tax. The responsibility for the payment of this tax lies with contractors and they in turn will invoice employers appropriately. If, after the base date of the contract, the supply of goods and services to the employer becomes exempt from VAT, because of a change in the tax legislation, the employer remains responsible for the payment of input tax to the contractor. The amount that would be payable to the contractor is the amount equal to the input tax on the sub-contractor/suppliers supply of goods or services which cannot be recovered by the contractor due to the employer's exemption. The contractor can recover VAT for the services of a sub-contractor ensuring that the ultimate user (the employer) is responsible for its payment.

Valuation and certificate forms

Except where employers have their own forms on which they require interim payments to be certified, it is normal practice for standard forms published by the professional bodies to be used. Valuation forms are published by the RICS for the use of quantity surveyors. These comprise the valuation shown in Example 28.1 and the statement of retention (which accompanies the valuation form as an appendix) shown in Example 28.2. RIBA Publications publishes forms for architects to use in connection with interim certificates. These are the interim certificate and the statement of retention. This can be filled in directly from the information given in the quantity surveyor's valuation and a completed specimen is shown in Example 28.3. All these forms are published in separate pads and each pad contains notes on their use. It will also be seen that the quantity surveyor's valuation form contains several notes which form a useful reminder of the contractual obligations. At the time of writing, the new forms (for use with the SBC) were not available from the RICS or the RIBA so the example given relate to the since superseded JCT 98 contract.

Valuation for JCT Standard Form of Contract (1998 Edition)

RICS

Surveyor Fussedon Knowles & Partners
Upper Market Street
Borchester, BC2 1HH

Works New School Library
Park Street
Fairbridge

Valuation No: 8
Date of Issue: 20:09:1999
Reference: RWP/03102

To Architect/Contract Administrator
Ivor Buck Associates
Prospects Drive
Fairbridge

Employer
Cosmeston Preparatory School
Fairbridge

Contractor
LEM Construction Ltd
Perry Road
Fairbridge

Contract Sum £ 506, 207.00

As at 15 September 1999 I/We have made, in accordance with the terms of the Contract, an Interim Valuation, the basis on which the amount shown as due has been calculated is clauses 30.2 and 30.4 of the Conditions of Contract, and report as follows:

Gross Valuation (excluding any work or material notified to me/us by the Architect/Contract Administrator in writing as not being in accordance with the Contract).	£	347, 260.00
Less total amount of Retention as attached Statement.	£	17, 362.00
		329, 898.00
Less total amount of Interim Certificates previously issued by the Architect/Contract Administrator up to and including Interim Certificate No. 7 and any advance payment due for reimbursement by the date given below for the issue of the next Certificate.	£	258, 133.00
Balance (in words) Seventy one thousand seven hundred & sixty five pounds	£	71, 765.00

Signature L. W. Pipe.

Surveyor ~~Tech RICS/ARICS~~/FRICS

SPECIMEN

Notes
(1) All the above amounts are exclusive of VAT.
(2) The balance stated is subject to any statutory deductions which the Employer may be obliged to make under the provisions of the Construction Industry Scheme where the Employer is classified as a 'Contractor' for the purposes of the relevant Act.
(3) It is assumed that the Architect/Contractor will:
(a) satisfy him or herself that there is no further work or material which is not in accordance with the Contract.
(b) notify Nominated Sub-Contractors of payments directed for them of Retention held by the Employer in accordance with clause 35.13
(c) satisfy him or herself that the previous payments directed for Nominated Sub-Contractors have been discharged in accordance with clause 35.13
(4) The Architect/Contract Administrator's Certificate should be issued on (see clause 30.2).

Example 28.1 Interim valuation.

Statement of Retention and of Nominated Sub-Contractors' Values

RICS

Surveyor: Fussedon Knowles & Partners, Upper Market Street, Borchester, BC2 1HH

Works: New School Library, Park Street, Fairbridge

This Statement relates to:
Valuation No: 8
Date of Issue: 20:09:1999
Reference: RWP/03102

	Gross Valuation	Basis of Gross Valuation (See note 1)	Amount Subject to: Full Retention of %	Half Retention of %	No Retention	Amount of Retention	Net Valuation	Amount Previously Certified	Balance
	£	Clause No.	£	£	£	£	£	£	£
Main Contractor	334,210.00		334,210.00	–	–	16,710.00	317,500.00	258,133.00	59,367.00
Nominated Sub-Contractors									
1. Speedweld (UK) Ltd (steelwork)	2,700.00	4.17	2,700.00	–	–	135.00	2,565.00	–	2,565.00
2. Ivor Short Ltd (electrical)	1,450.00	4.17	1,450.00	–	–	72.00	1,378.00	–	1,378.00
3. Fineline Ltd (architectural metalwork)	8,900.00	4.17	8,900.00	–	–	445.00	8,455.00	–	8,455.00
TOTAL	347,260.00		347,260.00	–	–	17362.00	329,898.00	258,133.00	71,765.00

Notes
(1) The basis of the gross valuation in respect of Nominated Sub-Contractors is
for interim payments – clause 4.17 of NSC/C
for final payments – sub-contracts based on a lump sum – (clause 3.1 of NSC/A) clause 4.23 of NSC/C
sub-contracts based on measurement – (clause 3.2 of NSC/A) clause 4.24 of NSC/C
(2) No account has been taken of any discounts for cash which the Contractor may be entitled if discharging the balance within 17 days of the issue of the Architect/Contract Administrator
(3) The sums stated are exclusive of VAT

Example 28.2 Statement of retention.

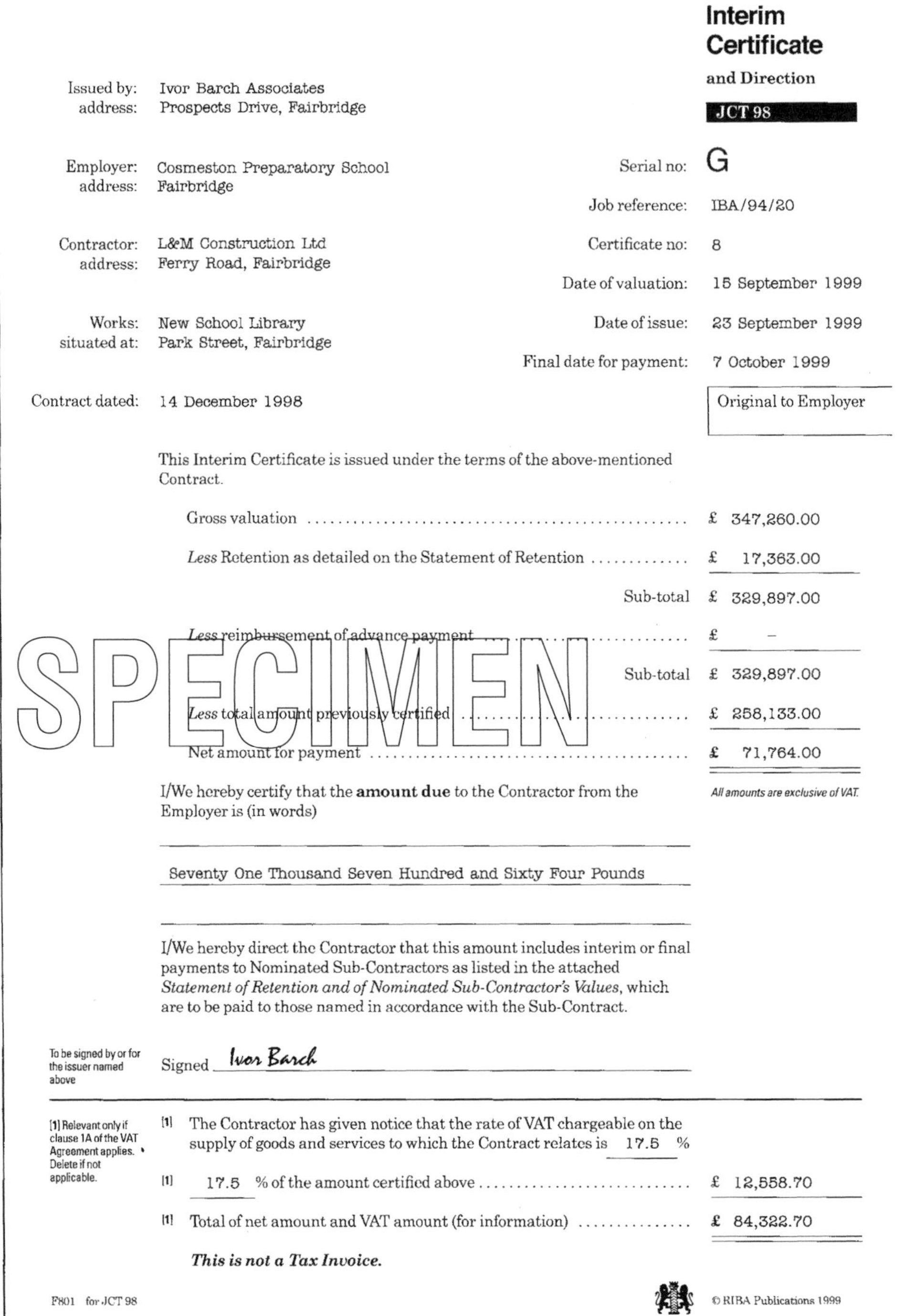

Interim Certificate

and Direction

JCT 98

Issued by: Ivor Barch Associates
address: Prospects Drive, Fairbridge

Employer: Cosmeston Preparatory School
address: Fairbridge

Serial no: G

Job reference: IBA/94/20

Contractor: L&M Construction Ltd
address: Ferry Road, Fairbridge

Certificate no: 8

Date of valuation: 15 September 1999

Works: New School Library
situated at: Park Street, Fairbridge

Date of issue: 23 September 1999

Final date for payment: 7 October 1999

Contract dated: 14 December 1998

Original to Employer

This Interim Certificate is issued under the terms of the above-mentioned Contract.

Gross valuation	£ 347,260.00
Less Retention as detailed on the Statement of Retention	£ 17,363.00
Sub-total	£ 329,897.00
Less reimbursement of advance payment	£ –
Sub-total	£ 329,897.00
Less total amount previously certified	£ 258,133.00
Net amount for payment	£ 71,764.00

SPECIMEN

I/We hereby certify that the **amount due** to the Contractor from the Employer is (in words)

All amounts are exclusive of VAT.

Seventy One Thousand Seven Hundred and Sixty Four Pounds

I/We hereby direct the Contractor that this amount includes interim or final payments to Nominated Sub-Contractors as listed in the attached *Statement of Retention and of Nominated Sub-Contractor's Values*, which are to be paid to those named in accordance with the Sub-Contract.

To be signed by or for the issuer named above

Signed Ivor Barch

[1] Relevant only if clause 1A of the VAT Agreement applies. Delete if not applicable.

[1] The Contractor has given notice that the rate of VAT chargeable on the supply of goods and services to which the Contract relates is 17.5 %

[1] 17.5 % of the amount certified above £ 12,558.70

[1] Total of net amount and VAT amount (for information) £ 84,322.70

This is not a Tax Invoice.

F801 for JCT 98

Example 28.3 Interim certificate.

Chapter 29
Completion, Defects and the Final Account

A building contract does not come to an end until the architect/contract administrator issues his final certificate and even then actions for breach of contract can be commenced within six or twelve years of the breach, depending upon whether the contract had been executed under hand or as a deed (see Chapter 22). Completion of the works can take place in three stages:

- practical completion
- sectional completion (if applicable) or partial possession
- completion of making good defects.

Practical completion

In the SBC practical completion of the works or section is defined by reference to clause 2.30. When in the opinion of the architect/contract administrator the contractor:

- has achieved practical completion of the works or section, including all authorised variations
- has provided such information as is reasonably required by the planning supervisor for the preparation of the health and safety file required by CDM94
- has supplied the employer with the required drawings and information, including any contractor's design documents, showing the building's maintenance and operation

then the architect/contract administrator is required to issue a certificate to that effect and practical completion of the works or section will, for the purposes of the contract, be deemed to have taken place on the day named in that certificate. This date is significant as it has an important influence on a number of conditions in the contract and disputes that may arise under it. RIBA Publications publishes a standard pro forma Certificate of Practical Completion and a completed copy of this form is given as Example 29.1 at the end of this chapter.

Over the years the courts have defined practical completion of the works in differing ways but the safest approach for any architect/contract administrator to adopt is to certify practical completion of the works or section only when every item of work has been satisfactorily completed. However, the architect/contract administrator is often put under pressure to certify practical completion prematurely to enable the employer to gain occupation of his building. The consequences to the employer and the contractor of issuing the certificate of practical completion whilst items of work remain incomplete, even if they are of a minor nature, must be considered by the architect/contract administrator – for example, the contractor will have to complete work in an occupied (and perhaps otherwise finished) building, giving rise to additional problems related to health, safety and security. If an architect/contract administrator were minded to issue the certificate of practical completion when items of work remain incomplete, he would be well advised separately to identify, in writing to the contractor, the items of work that remain to be completed, an order of priority for their completion and the date(s) by when they are to be completed. Practical completion ought *never* to be certified when there is any outstanding *defective* work (as distinct from incomplete work).

The achievement of practical completion is used in the SBC as the datum for the effective commencement or cessation of various provisions. These are:

Clause	**Description**
2.4	The contractor retains possession of the site and the works or section up to and including the date of issue of the certificate of practical completion and the employer is not entitled to take possession of the works or section until that date, except as provided for in clause 2.33 (see later in this chapter).
2.28.2	Not later than the expiry of 12 weeks after the date of practical completion the architect/contract administrator has to fix a completion date later or earlier than previously fixed or confirm the completion date previously fixed. Nevertheless, an architect/contract administrator cannot fix an earlier completion date than the original contract completion date. Nor can he fix a completion date earlier than the date subject to a 'pre-agreed adjustment' under a Schedule 2 Quotation, unless work that is the subject of that quotation has been omitted.
2.32	The employer is entitled to recover liquidated and ascertained damages from the contractor in respect of the period between the completion date and the date of practical completion if a non-completion certificate has been issued in accordance with 2.31.
2.38	The rectification period (six months unless the contract particulars state to the contrary) runs from the day named in the certificate of practical completion of the works or section.
2.38	Defects, shrinkages or other faults due to materials or workmanship not being in accordance with the contract, or failure by the contractor to

comply with the contractor's designed portion, which occur following practical completion of the works or section, have to be made good by the contractor at no cost to the employer.

2.40 The contractor has to supply the employer with the required drawings and information showing and/or describing any contractor's designed portion work as built and, where relevant, its maintenance and operation, before the date of practical completion.

3.6 The contractor is wholly responsible for carrying out and completing the works in accordance with the contract.

4.5 Not later than six months after issue of the practical completion certificate, the contractor has to provide to the architect/contract administrator or to the quantity surveyor, all documents necessary for the purposes of the adjustment of the contract sum.

4.9.2 Interim certificates have to be issued on the dates provided in the contract particulars up to the date of practical completion or to within one month after it. (Thereafter interim certificates have to be issued as and when further amounts are ascertained as payable to the contractor and upon whichever is later of expiry of the rectification period – itself calculated by reference to the date of practical completion – or the issue of certificate of making good.)

4.20 In interim certificates the full retention percentage may be deducted from the value of work which has not reached practical completion, but only half the retention percentage may be deducted from the value of work which has reached practical completion (until such time as the certificate of making good defects is issued under clause 2.39).

6.5.1 Where any part of the works or sections are not subject of a certificate of practical completion, injury or damage to the works is excluded from any insurance required pursuant to clauses 6.5.1.1 to 6.5.1.9 (insurance – liability, etc. of employer).

6.7 The joint names insurance policies required by the applicable clause(s) have to be maintained by either the contractor or the employer, as the case may be, up to and including the date of issue of the certificate of practical completion (or the date of termination of the contractor's employment if that is earlier). Upon deemed practical completion of a relevant part, in the case of vacant possession, the employer is responsible for insuring the completed part of the works.

7.2 Where the contract particulars state that this clause applies the employer may, at any time after practical completion of the works or section, assign to a transferee or lessee the right to bring proceedings in the name of the employer to enforce any of the terms of the contract made for the benefit of the employer.

8.4 If the contractor makes a specified default prior to the date of practical completion then, subject to giving the appropriate notices, the employer is entitled to terminate the employment of the contractor.

8.9 If the carrying out of the works, or substantially the whole of the uncompleted works, is suspended for a continuous period (of a length stated in the contract particulars) or the employer makes a specified default, prior to the date of practical completion, then, subject to giving the appropriate notices, the contractor is entitled to terminate his employment.

8.11 If the carrying out of the works, or substantially the whole of the uncompleted works, is suspended for a continuous period (of a length stated in the contract particulars), prior to the date of practical completion, for one or more of the specified reasons then, subject to giving the appropriate notices, either party is entitled to terminate the contractor's employment.

If the project is large, it is quite feasible to have the situation where a certificate of practical completion is issued in respect of one section before work commences on other sections. Similarly the certificate of making good defects can be issued in respect of one section whilst the work remains incomplete on other sections.

Partial possession

As previously noted, sectional completion arises when the employer requires the works to be completed in phased sections. Partial possession, however, arises out of a post-contract agreement between the employer and the contractor as provided for in SBC clauses 2.33 to 2.37. If, before practical completion, the employer, with the consent of the contractor, takes possession of a part of the works or a section, clauses 2.33 to 2.37 apply with the following effects:

- The architect/contract administrator has to immediately issue, to the contractor, a written statement identifying the part of the works or section taken into possession (the relevant part) and the date on which that possession occurred (the relevant date).
- In respect of the relevant part, practical completion is deemed to have occurred and the rectification period is deemed to have commenced on the relevant date.
- The architect/contract administrator has to issue a certificate of making good defects in respect of the relevant part when any defects, shrinkages or other faults in the relevant part that were required to be made good have been made good.
- The contractor and the employer are relieved of their respective duties to insure the relevant part under clause 6.7 and Opion A, Option B or Option C, whichever is applicable, but when Option C is applicable the employer will be obliged to insure the relevant part from the relevant date.
- The rate of liquidated damages (LD) stated in the contract particulars has to be proportionally revised as follows:

$$Revised\ LD = \frac{Original\ LD \times (Contract\ sum - Value\ of\ Relevant\ Part\ in\ Contract\ sum)}{Contract\ sum}$$

Possession of the building

A short time before practical completion the architect/contract administrator should ensure that the contractor has removed all plant, surplus materials, rubbish and temporary works from the site. When practical completion has been achieved and the architect/contract administrator has issued the certificate to that effect, the contractor ceases to be responsible for the works or section of the site and relinquishes possession of them to the employer. At this time the employer becomes responsible for all insurance matters relating to the site, the building(s) and contents.

At this stage it is advisable for the design team and the contractor to have a 'hand-over' meeting with the employer. (See also Chapter 24 on meetings.) During this meeting:

- the contractor can formally hand over to the employer all keys, properly labelled
- the contractor can hand over to the employer the building log-book, referred to in the Building Regulations 2000 Part L2 Section 3, giving details of the installed building services plant and controls, the method of operation and maintenance, and other details that collectively enable energy consumption to be monitored and controlled
- the employer and/or his staff can be fully briefed on the operation of the building and its services (when the building and/or its services are complex this exercise may involve special training sessions for those who are to operate and maintain the facilities)
- the contractor can hand over to the employer all relevant operating and maintenance manuals, guarantees and, where appropriate, a supply of spare parts for maintenance of critical equipment, etc.
- the design team and/or the contractor can hand over to the employer a full set of 'as built' drawings
- the planning supervisor can hand over to the employer the completed health and safety file.

Defects and making good

SBC clause 2.38 requires that not later than 14 days after the expiry of the rectification period the architect/contract administrator has to prepare and deliver to the contractor as an instruction a schedule of all defects, shrinkages or other faults which are due to materials or workmanship not in accordance with the contract that have appeared during the rectification period.

In practice it is usual for the employer to keep a record of such defects as they occur or are noticed and for this record to form the basis of the architect/contract administrator's schedule, which would normally be compiled by the design team during a thorough inspection of the works at the end of the rectification period. Under normal circumstances the contractor has to make good all items on the schedule of defects at no cost to the employer and within a reasonable time of having received the instruction. However, SBC clause 2.38 does allow the architect/contract administrator, with the employer's consent, to instruct the contractor not to make good some or all of the scheduled items. When the architect/contract administrator does so instruct, an appropriate deduction is made from the contract sum in respect of those items not to be made good.

If the architect/contract administrator requires any defect, shrinkage or fault to be made good during the rectification, SBC clause 2.38 permits him to so instruct the contractor, who has to comply with the instruction within a reasonable time. If the contractor fails to make good any item within a reasonable period the architect/contract administrator may, after having given the contractor the notice required by SBC clause 3.11, employ and pay others to execute the work and deduct all costs thereby incurred from monies due to the contractor. If there is insufficient money due to the contractor fully to defray the costs incurred, the shortfall is recoverable by the employer from the contractor as a debt.

When the architect/contract administrator is satisfied that all defects, shrinkages and other faults that were formally required to be made good have been made good, a certificate to that effect must be issued. RIBA Publications has produced a pro forma certificate of completion of making good defects and a completed copy is given as Example 29.2 at the end of this chapter.

Final account

The responsibilities of the contractor, the architect/contract administrator and the quantity surveyor in connection with the final account are set out at the beginning of SBC clause 4.5 and are:

- not later than six months after practical completion of the works the contractor has to send to the architect/contract administrator (or the quantity surveyor if so instructed by the architect/contract administrator) all documents necessary for the adjustment of the contract sum
- within three months of the contractor sending all necessary documents, the architect/contract administrator (or the quantity surveyor if so instructed by the architect/contract administrator), has to ascertain any outstanding loss and/or expense and the quantity surveyor has to prepare a statement of all other adjustments to be made to the contract sum – the ascertainment of loss and/or expense and the statement of adjustments are, together, commonly referred to as the final account

Final account

- when complete, the architect/contract administrator has to send a copy of the final account to the contractor.

It is worth noting that, under the SBC, the architect/contract administrator and the quantity surveyor have unilateral responsibility for the preparation of the final account – there is no requirement for them to obtain the contractor's agreement to their ascertainment and adjustments. However, in practice, it is normal for the consultants and the contractor to co-operate in reaching an agreed final account. When such agreement cannot be reached the contractor's only recourse is to invoke the contract dispute resolution procedures.

Adjustment of the contract sum

SBC clause 4.3 also sets out all of the matters that must be dealt with in the final account in order to properly adjust the contract sum in accordance with the conditions of contract. These are summarised below.

To be deducted or added as the case may be

- the amount of any valuation agreed by the employer and the contractor for a variation

- the value of any Schedule 2 quotation for a variation for which the architect/contract administrator has issued a confirmed acceptance
- the value of any variation in the premium for renewing terrorism cover under the joint names policy, when the contractor insures the works under insurance Option A.

To be deducted

- prime cost sums and any contractor's profit thereon included in the contract bills – this appears to be a throwback to JCT 98 since there are no longer nominated sub-contractors to which prime cost sums relate, moreover the 'to be added' element of the SBC is silent in prime cost sums
- all provisional sums and the value of all work included by way of approximate quantities in the contract bills
- the amount of the valuation of variations which are omissions
- the amount included in the contract bills for work which, due to a variation, has to be executed under substantially changed conditions
- the amount included in the CDP analysis which, due to a variation, has been omitted from the employer's requirements or has to be executed under substantially changed conditions
- any amount in respect of errors in setting out which the architect/contract administrator instructs are not to be amended
- any amount in respect of work not in accordance with the contract which the architect/contract administrator allows to remain
- any amount in respect of any defects, shrinkages or other faults which the architect/contract administrator instructs are not to be made good
- any amount allowable to the employer for whichever fluctuation option applies
- any other amount which is required, by the contract, to be deducted from the contract sum.

To be added

- any amount payable by the employer in respect of statutory fees and charges legally demandable under any of the statutory requirements
- any amount payable by the employer for royalties arising out of compliance with an architect/contract administrator's instruction
- any amount payable by the employer for opening up work for inspection and testing
- any amount payable by the employer to the contractor for complying with an architect/contract administrator's instruction to take out a policy of insurance under clause 6.5.1
- the amount of the valuation of variations, other than those which are omissions

- the amount of the valuation of any work, including CDP work, which due to a variation has to be executed under substantially changed conditions
- the amount of the valuation of work executed as the result of instructions relating to the expenditure of provisional sums and all work included by way of approximate quantities in the contract bills
- any amount ascertained in respect of loss and/or expense arising from matters materially affecting the contractor's regular progress (of the works) or from the discovery of antiquities
- the amount of any costs incurred by the contractor if the employer defaults under insurance Option B or Option C, or under Option A where an additional premium is required by the insurers due to the employer's early use of the site, works or part of them under clause 2.6.1
- any amount payable to the contractor for whichever fluctuation option applies
- any other amount which is required, by the contract, to be added to the contract sum.

It is good practice for each of the foregoing items to be identified separately in the final account and for separate amounts to be given for each variation.

Practical considerations

Whenever measurements have to be taken for the purpose of valuation the contractor must be given the opportunity of being present and of taking whatever notes and measurements he may require. It is to be noted that this facility is to be extended to the contractor both when measurements are to be taken on site, directly from the work, and when they are to be taken from drawings and other documents.

It is a worthwhile exercise to draft the final account as construction proceeds as it can be of considerable use to the building team for the purposes of:

- post-contract cost control and the preparation of regular financial reviews
- valuations for interim certificates
- speeding up the production of the final account once practical completion of the works has been achieved.

Whether or not this is done the design team must bear in mind the SBC requirement to complete the final account within three months of receipt from the contractor of all the documents necessary for its production. Any delays in doing so are likely to cause either the contractor or the employer to incur additional financing costs.

Finally, clause 4.2 states that any adjustments to the contract sum shall only be carried out in accordance with the express provisions of the contract. Despite this requirement and the architect/contract administrator's unilateral responsibility, or the quantity surveyor's if so instructed, it is the adjustment of the contract sum and the subsequent final account statement which is frequently disputed by contractors.

In some instances, and contrary to clause 4.2, the employer and contractor may pragmatically choose to agree a 'commercial' settlement. Should a commercial settlement be negotiated and agreed, this should be recorded in a separate agreement, which is itself a contract.

Final certificate

SBC clause 4.15 requires the architect/contract administrator to issue the final certificate within two months of whichever of the following events occurs last:

- the end of the rectification period in respect of the works or, where there are sections, expiry of the period for the last section; or
- the date of issue of the certificate of making good in respect of the works or, where there are sections, issue of the last certificate of making good; or
- the date on which the architect/contract administrator sends a copy of the final account to the contractor.

The amount of the balance due either from the employer to the contractor or vice versa expressed in the final certificate is the difference between the total of the final account and the amounts previously stated as due in interim certificates plus the amount of any advance payments paid. The same clause also preserves the contractor's rights to any amounts previously included in interim certificates that have not been paid by the employer. RIBA Publications publishes a pro forma final certificate – a completed copy is given as Example 29.3 at the end of this chapter.

SBC clause 1.10 prescribes the intended effect of the final certificate (if not contested), which can be summarised as:

- conclusive evidence that where the particular quality of any materials etc. or standard of any workmanship is to be for the approval of the architect/contract administrator, such items are to the reasonable satisfaction of the architect/contract administrator
- conclusive evidence, save for accidental errors, that the contract sum has been properly adjusted in accordance with all the terms of the contract
- conclusive evidence that only those extensions of time that are due under the contract have been given
- conclusive evidence that the reimbursement of loss and/or expense is in full and final settlement of all and any claims that the contractor may have in respect of matters affecting the regular progress, whether breach of contract, duty of care, statutory duty or otherwise.

Note: In the examples which follow, it is to be noted that, at the time of writing, the new forms (for use with the SBC) were not available from the RIBA so the examples given relate to the since superseded JCT 98 contract.

Certificate of
Practical Completion

JCT 98 / IFC 98

Issued by: Ivor Barch Associates
address: Prospects Drive, Fairbridge

Employer: Cosmeston Preparatory School
address: Fairbridge

Job reference: IBA/94/20

Contractor: L&M Construction Ltd
address: Ferry Road, Fairbridge

Certificate no: 1

Issue date: 28 January 2000

Works: New School Library
situated at: Park Street, Fairbridge

Contract dated: 14 December 1998

Under the terms of the above-mentioned Contract,

I/we hereby certify that in my/our opinion

Practical Completion of

* the Works

*Delete as appropriate

~~* Section no. of the Works~~

has been achieved

* and the Contractor has complied with the contractual requirements in respect of information for the health and safety file

This item applies to JCT 98 only.

* and the Contractor has supplied the ~~specified drawings and information relating to Performance Specified Work~~

on 28 January 20 00

To be signed by or for the issuer named above

Signed *Ivor Barch*

Distribution			
☐ Employer	☐ Structural Engineer	☐ Planning Supervisor	☐
☐ Contractor	☐ M&E Consultant	☐	☐
☐ Quantity Surveyor	☐ Clerk of Works	☐	☐ File

F853A/B for JCT 98 / IFC 98

Example 29.1 Certificate of practical completion.

Certificate of
Completion of

Making Good Defects

JCT 98

Issued by: Ivor Barch Associates
address: Prospects Drive, Fairbridge

Employer: Cosmeston Preparatory School
address: Fairbridge

Job reference: IBA/94/20

Contractor: L&M Construction Ltd
address: Ferry Road, Fairbridge

Certificate no: 1

Issue date: 8 September 2000

Works: New School Library
situated at: Park Street, Fairbridge

Contract dated: 14 December 1998

Under the terms of the above-mentioned Contract,

I/we hereby certify that the making good of

any defects, shrinkages or other faults specified in a schedule of defects delivered to the Contractor as an instruction

and any other defect, shrinkage or fault which has appeared within the Defects Liability Period and has been required by an instruction to be made good

and relating to

*Delete as appropriate

*the Works referred to in the Certificate of Practical Completion

no. 1 dated 28 January 2000

~~*Section no. of the Works referred to in the Certificate of Practical Completion no. dated~~

* the part of the Works identified in the Statement of Partial Possession by the Employer

no. 1 dated 26 November 1999

was in my/our opinion completed on

6 September 20 00

To be signed by or for the issuer named above

Signed *Ivor Barch*

Distribution

- [ ] Employer
- [ ] Contractor
- [ ] Quantity Surveyor
- [ ] Structural Engineer
- [ ] M&E Consultant
- [ ] Clerk of Works
- [ ] Planning Supervisor
- [ ] File

F807A for JCT 98

Example 29.2 Certificate of making good defects.

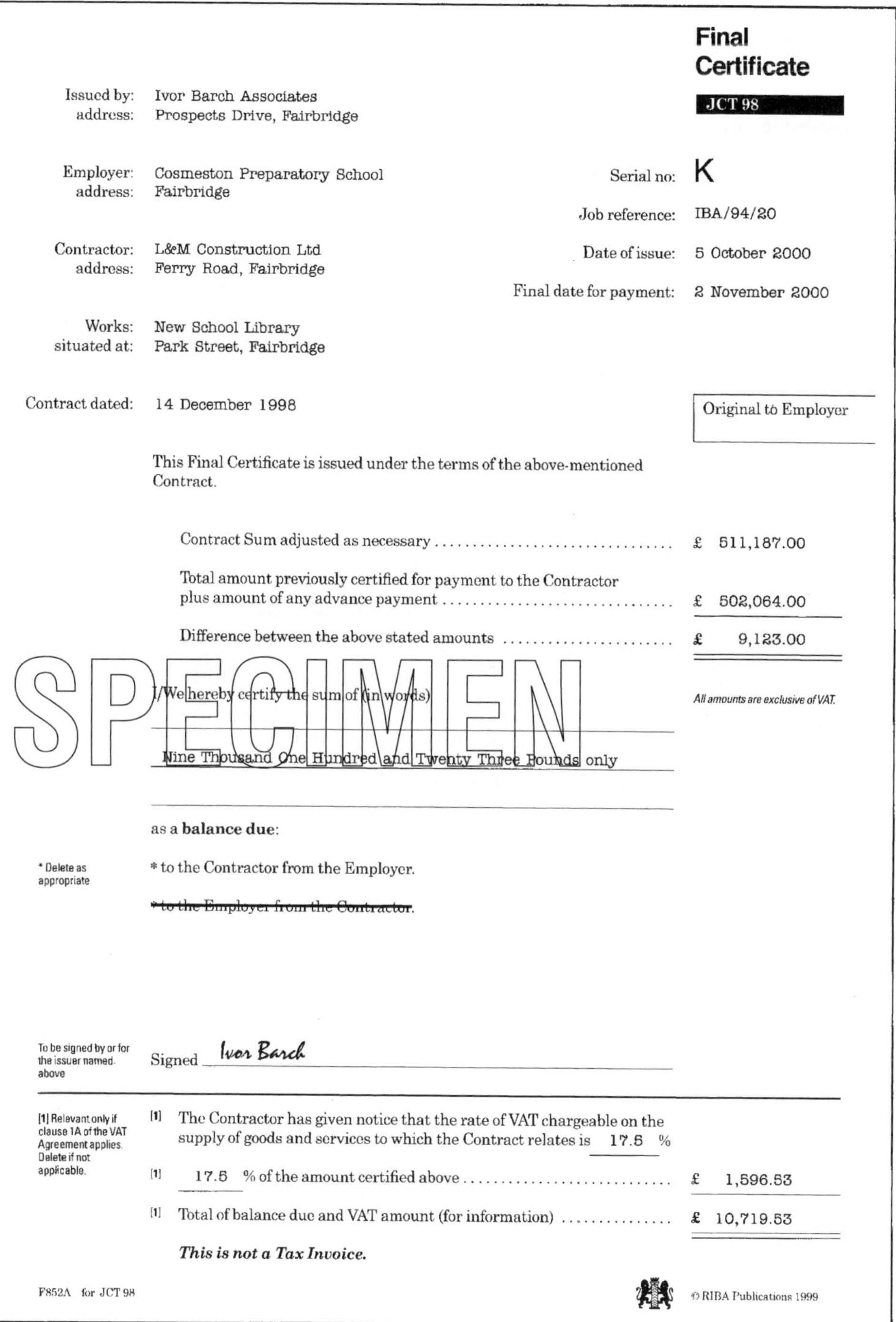

Final Certificate

JCT 98

Issued by: Ivor Barch Associates
address: Prospects Drive, Fairbridge

Employer: Cosmeston Preparatory School
address: Fairbridge

Serial no: K

Job reference: IBA/94/20

Contractor: L&M Construction Ltd
address: Ferry Road, Fairbridge

Date of issue: 5 October 2000

Final date for payment: 2 November 2000

Works: New School Library
situated at: Park Street, Fairbridge

Contract dated: 14 December 1998

Original to Employer

This Final Certificate is issued under the terms of the above-mentioned Contract.

Contract Sum adjusted as necessary .... £ 511,187.00

Total amount previously certified for payment to the Contractor plus amount of any advance payment .... £ 502,064.00

Difference between the above stated amounts .... £ 9,123.00

All amounts are exclusive of VAT.

SPECIMEN

I/We hereby certify the sum of (in words)

Nine Thousand One Hundred and Twenty Three Pounds only

as a **balance due**:

* Delete as appropriate

* to the Contractor from the Employer.

~~* to the Employer from the Contractor.~~

To be signed by or for the issuer named above

Signed *Ivor Barch*

[1] Relevant only if clause 1A of the VAT Agreement applies. Delete if not applicable.

[1] The Contractor has given notice that the rate of VAT chargeable on the supply of goods and services to which the Contract relates is 17.5 %

[1] 17.5 % of the amount certified above .... £ 1,596.53

[1] Total of balance due and VAT amount (for information) .... £ 10,719.53

This is not a Tax Invoice.

F852A for JCT 98

© RIBA Publications 1999

Example 29.3 Final certificate.

Chapter 30
Delays and Disputes

Introduction

Delays to the progress of the works and the loss and/or expense associated with such delays are perhaps the most common causes of dispute encountered in the administration of building contracts. It is not surprising, therefore, that the draughtsmen of the SBC paid particular attention to the procedures to be followed when delays occur, or are foreseen, as a result of which the contract period might become extended and/or the contractor may become entitled to the reimbursement of loss and/or expense.

If the employer (or those for whom he is responsible), through default, prevents the contractor from completing the works by the contracted completion date and there are no express contractual provisions enabling the employer to set a new date for completion, the contractor's obligation is transformed from one of having to complete by the contracted date to one of completing in a reasonable time. In other words, time is 'at large' and employers lose any right that they may otherwise have to levy liquidated damages. It may be seen, therefore, that, far from favouring only contractors, the existence and proper execution of an extension of time clause within a contract is of the utmost importance to employers in that it preserves their other contractual rights should they or their representatives cause the works to be delayed by default.

All delays will fall into one of the following three categories:

- delays caused by the contractor
- delays caused by the employer or his representatives
- delays caused by events outside the control of both the contractor and the employer (often referred to in the industry as 'shared risk events' or 'neutral events').

Although only delays caused by employers or their representatives or by neutral events are recognised by SBC clause 2.29 as relevant events which may give rise to an extension of time, clause 2.27.1 requires the contractor to give written notice to the architect/contract administrator whenever it becomes reasonably apparent that the progress of the works or any section is being or is likely to be delayed, irrespective of cause.

Delays caused by the contractor

Within this category are all those delays that could be avoided if the contractor proceeds regularly and diligently with the works and uses best endeavours at all times to prevent delays. The SBC effectively requires contractors to take all measures within their control to ensure:

- that there is an adequate labour force on the job
- that the necessary goods and materials are on site whenever they are needed
- that the works are not delayed by sub-contractors.

If the contractor does not take all such measures and delays occur, there is no provision in the SBC to enable the architect/contract administrator to grant an extension of time in respect of them. The responsibility for any such delays lies with the contractor and under SBC clause 2.32 he may well become liable to the employer for liquidated damages in respect of the delay period.

If the contractor fails to proceed regularly and diligently with the works, the employer's ultimate sanction, subject to giving the required notices, is to terminate the contractor's employment under SBC clause 8.4.

Delays caused by employers or their representatives

Certain of the relevant events recognised by SBC clause 2.29 result from action or inaction on the part of the employer or his representatives. These are:

- compliance with the architect/contract administrator's instructions to open up work for inspection or testing unless the test or inspection shows that the work is not in accordance with the contract
- compliance with the architect/contract administrator's instructions in regard to the expenditure of undefined provisional sums
- compliance with an architect/contract administrator's instruction regarding antiquities
- deferment of giving possession of the site or any section of the works
- compliance with the architect/contract administrator's instructions requiring a variation, except those for which a Schedule 2 quotation has been accepted
- the execution of work for which an approximate quantity is included in the contract bills which is not a reasonably accurate forecast of the quantity of work required

- compliance with the architect/contract administrator's instructions in regard to any discrepancies within the contract documents
- compliance with the architect/contract administrator's instructions in regard to the postponement of any work
- suspension by the contractor of the performance of his obligations under the contract as a result of the employer's failure to pay a due amount in full
- any impediment, prevention or default, whether by act or omission by the employer, the architect/contract administrator, the quantity surveyor or any of the employer's personnel except to the extent that it was caused or contributed to by any default, whether act or omission, of the contractor or any of the contractor's personnel.

The final bullet point is a catch-all 'relevant event'. As Chapter 23 has shown, a significant number of the employer's obligations under the contract are carried out by the design team and many of these are to assist contractors in complying with their own obligations. As an example, the architect/contract administrator's failure to issue design information in accordance with the information release schedule would fall within 'any impediment, prevention or default' by the architect/contract administrator for whom the employer is responsible.

Not only do all delays within this category entitle the contractor to an extension of time if they are likely to delay the completion of the works beyond the completion date; they may also cause the contractor to suffer loss and/or expense which, if incurred and subject to the issue of the relevant notices, is reimbursable under the provisions of SBC clause 4.23.

Further, if an employer fails to comply with CDM94 and/or if the carrying out of the works is suspended for a continuous period of the length stated in the contract particulars to the SBC by reason of one or more of the last four relevant events listed above (including the 'catch all' one), or by reason of compliance with a specified architect/contract administrator's instruction requiring a variation, contractors have the ultimate sanction under SBC clause 8.9 (subject to giving the required notices) of terminating their employment.

Delays caused by events outside the control of either party

There are occasions when delays arise due to circumstances over which neither the employer nor the contractor has any control. These neutral events, which are also recognised by SBC clause 2.29, are:

- a statutory undertaker executing, or failing to execute, work in pursuance of its statutory obligations in relation to the works
- exceptionally adverse weather conditions

- loss or damage caused by fire, lightning, explosion, storm, tempest, flood, bursting or overflowing of water tanks, apparatus or pipes, earthquake, aircraft and other aerial devices or articles dropped therefrom, riot or civil commotion
- local combination of workmen, strike or lockout affecting any of the trades employed on the works or any of the trades engaged in the preparation, manufacture or transportation of any of the goods or materials required for the works
- civil commotion or the use or threat of terrorism and/or the activity of the relevant authority in dealing with such use or threat
- compliance with the architect/contract administrator's instructions in regard to what is to be done with any fossils, antiquities and other objects of interest or value that may be found on the site
- exercise by the UK government of any statutory power which directly affects the execution of the works
- force majeure
- the deferment by the employer of giving possession of the site when clause 2.5 applies.

All delays within this category entitle the contractor to an extension of time if they are likely to delay the completion of the works beyond the completion date, but with the exception of those caused by compliance with the architect/contract administrator's instructions in regard to what is to be done with any fossils, antiquities and other objects of interest or value that may be found on the site, they do not entitle the contractor to reimbursement of any loss and/or expense incurred. In this way the parties share the burden of any delays caused by these neutral events. Not being a matter attributable to fault on the part of employers or their teams, reimbursement of any direct loss and/or expense arising from architect's instructions in regard to what is done with any fossils, anti-quities etc. is covered by SBC clause 3.24 and not clause 4.23.

It is to be noted that in the case of delays, the SBC imposes an obligation on the contractor to take all practicable steps to avoid or reduce the delay. Should such delays occur, the architect/contract administrator needs to be satisfied that the contractor has taken all such steps before consideration is given to granting an extension of time.

Differences in the interpretation of certain of these neutral relevant events may give rise to disputes. It is therefore worth considering them in more detail.

Force majeure

This phrase comes from French law (Code Napoléon) and does not have a precise definition in English. Its literal translation is 'superior force' and in its broadest usage examples would include acts of God, war, strikes, weather conditions, epidemics, direct legislative or administrative interference, etc. However, several of these matters are provided for elsewhere in the SBC, hence force majeure is included as a general catch-all term to provide for any other event or effect that cannot reasonably be anticipated or controlled and which is not otherwise expressly included in the contract.

Exceptionally adverse weather conditions

In earlier editions of the JCT standard forms of building contract the term 'exceptionally inclement weather' was used. The present term was introduced in the 1980 editions of the JCT contracts following the unusually long, hot summer of 1976, when many building projects were delayed by exceptionally adverse, but by no means inclement, weather.

'Exceptionally' is clearly the important word in this phrase and must be interpreted according to the time of year and the timing of the project anticipated by the contract documents. Thus if it were known at the time the contract were let that the execution of weather-sensitive work would span the winter period, then if the works were delayed by a two-week period of snow and frost during January such weather would most probably not be regarded as exceptionally adverse and would therefore not be a relevant event. However, should the snow and frost persist for a continuous period of eight weeks, this may well be regarded as being exceptionally adverse and therefore constitute a relevant event giving entitlement to an extension of time.

It is also important when interpreting the word 'exceptionally' to have regard to the location of the works, for what may be exceptional in the home counties may not be so in the north of England.

SBC procedure in the event of delay

Clearly, it is in the interests of both the employer and the contractor to take all possible steps to avoid delays occurring, but when they do architects/contract administrators must bear in mind that they may give extensions of time only in respect of delays caused by the relevant events listed in clause 2.29 and they may certify payment of loss and/or expense arising only from the relevant matters listed in clauses 4.24 and 3.24. Any other delays or loss and/or expense that may be caused by the employer cannot be dealt with by the architect/contract administrator.

It is worth noting that any delays caused by default of the employer that cannot be dealt with under the contract, for example failure to give possession of the site within the period of deferment stated in the contract particulars, can result in time being 'at large' (see above) and employers losing any right that they may otherwise have to levy liquidated damages.

Best endeavours

SBC clause 2.28.6.1 requires the contractor to 'use constantly his best endeavours to prevent delay in the progress of the Works and to prevent the completion of the

Works being delayed or further delayed beyond the Completion Date' as a condition precedent to the architect/contract administrator granting an extension of time. This is a demanding proviso and one which would seem to impose on the contractor an obligation to do everything possible to achieve the earlier completion date. It has been suggested that this obligation does not contemplate the expenditure of substantial sums of money. However, in using their best endeavours, contractors are expected to employ the same resources as would a prudent person to achieve his own ends and in doing so they will incur some cost. It is suggested that if the cause of delay is a relevant matter under clause 4.24 the reasonable cost to contractors of using best endeavours is reimbursable as part of their loss and/or expense – otherwise it is not a recoverable cost. When the cost is not recoverable contractors will have to decide the extent of the measures they are prepared to take in order to reduce their exposure to liquidated damages in respect of any delays they may have caused.

Notification of delay

Whenever it becomes apparent that the progress of the works or any section is being or is likely to be delayed clause 2.27 requires contractors to give written notice to the architect/contract administrator of the pertinent conditions and the event(s) causing delay. It also requires them, either in that notice or in a further written notice issued as soon after the first as possible, to identify those events that in their opinion are relevant and to give particulars of the expected effects and an estimate of the expected delay in the completion of the works or section in respect of each such event. These particulars and estimates have to be updated whenever necessary or as required by the architect/contract administrator.

If the architect/contract administrator is of the opinion that any of the notified events causing a delay is a relevant event and that the completion of the works or section is likely to be delayed beyond the completion date by reason of that delay, he has to fix a later completion date and notify the contractor accordingly in writing. RIBA Publications has published a pro forma notification of a revised completion date – a completed copy is given as Example 30.1 at the end of this chapter (it is to be noted that, at the time of writing, the RIBA's form for use with the SBC was not available so the example given instead relates to the since superseded JCT 98).

New completion dates

The architect/contract administrator should fix a new completion date as soon as it is reasonably practicable to do so and within 12 weeks of receiving the required notice from the contractor or, where the period from receipt to the contractual completion date is less than 12 weeks, endeavour to do so by the completion date. In the event that the architect/contract administrator fixes a new completion date, he must state which of the relevant events he has taken into account and the extent, if any, to which he has had regard to variations requiring an omission of work.

When the completion date is adjusted for the first time it can only be extended to a later date. However, after this has first happened, if the architect/contract administrator issues any instructions which result in the omission of work by way of:

- a variation and/or
- the omission of a provisional sum for defined work

then a new earlier completion date may be fixed, provided that, having regard to the work omitted, it is fair and reasonable to do so. However, when fixing an earlier completion date the architect/contract administrator is not allowed to alter the length of any extension of time required by the contractor and for which a Schedule 2 quotation has been accepted, nor can a new completion date be fixed that is earlier than the date for completion stated in the contract particulars.

Final adjustment

Not later than 12 weeks after the date of practical completion, the architect/contract administrator is required to finalise the position regarding the completion date by taking one of the following three courses of action. Whichever course of action the architect/contract administrator takes, he must commit his decision to writing and forward it to the contractor.

Fix a completion date later than that previously fixed

If, upon reviewing all previous decisions regarding extensions of time and upon reconsideration of all relevant events (whether or not specifically notified by the contractor) the architect/contract administrator reaches the conclusion that it is fair and reasonable to fix a completion date later than that previously fixed, he may do so. It will thus be seen that due notification by the contractor of a relevant event is not, in the end, a condition precedent to a retrospective grant of an extension of time, although it is a condition precedent to seeking an extension of time before practical completion.

Fix a completion date earlier than that previously fixed

It is to be noted that fixing an earlier completion date has to be related to the omission of work – it cannot result from the architect/contract administrator reducing previously granted extensions of time which, with the benefit of hindsight, are thought to be overly generous. As when fixing an earlier completion date prior to practical completion, the architect/contract administrator is not allowed to alter the length of any extension of time required by the contractor in respect of a Schedule 2 quotation, nor is he allowed to fix a new completion date that is earlier than the date for completion stated in the contract particulars.

Confirm the completion date previously fixed

After the final review it may be that no adjustment is necessary and the contractual completion date for the purposes of calculating any liquidated damages (see below) and the rectification period can remain either as stated originally in the contract particulars or as adjusted following a relevant event prescribed in clause 2.29.

Duties and decisions

There is little doubt that these procedures impose upon the contractor and the architect/contract administrator not inconsiderable obligations whenever there is a likelihood of delay. When seeking an extension of time the contractor must be specific and the architect/contract administrator must deal with the matter as and when it arises. The intention of the SBC is quite clear – decisions on extensions of time are to be made as quickly as possible after the occurrence of a relevant event that is likely to cause a delay. There is no place under the SBC for the practice of waiting to see what the delay actually is at the end of the contract and then arguing about whether or not an extension of time is justified. Such practices, if followed, may well cause the contractor to incur additional costs in accelerating the works to meet an unnecessarily onerous completion date.

Whenever the architect/contract administrator has to review the completion date it should not be overlooked that the master programme, which the contractor is obliged to provide, may well contain much valuable information to assist in the deliberations.

Perhaps the greatest difficulty that the architect/contract administrator is likely to encounter in relation to the procedures to be followed in the event of delay is in deciding whether or not the notice, particulars and estimates that the contractor is required to provide are sufficient to make it practicable for a new completion date to be fixed. In making this decision the architect/contract administrator should, as always, act fairly and must avoid the use of contrived allegations of insufficiency of notice, particulars or estimate as an excuse for delaying the fixing of a new completion date.

Reimbursement of loss and/or expense under the SBC

Any disruption or delay caused by default of the employer (or those for whom he is responsible) would ordinarily give the contractor a right to claim damages at common law. As clause 4.23 sets down the procedures to be followed in the event of matters materially affecting the relevant progress (of the works or section) and clause 4.24 provides a fairly comprehensive list of relevant matters to recompense the contractor for losses arising from the employer's acts or defaults, it could be suggested that this restricts the contractor's common law entitlement if it were not for the provisions of clause 4.6 that states: 'The provisions of clauses 4.23 to 4.25 are without prejudice to any other rights or remedies which the Contractor may possess'.

If the regular progress of the works is materially affected due to deferment of giving possession of the site (where clause 2.5 is applicable) or by one or more of the relevant matters listed in clause 4.24, then subject to the timely issue of the relevant notices, the contractor is entitled to be reimbursed any direct loss and/or expense that may thereby be incurred and which is not reimbursable by a payment under any other provision in the contract. These relevant matters are, in fact, the same as the relevant events, recognised by clause 2.29, resulting from the actions or inaction of employers or their representatives and which are listed earlier in this chapter.

If contractors consider that they have incurred or are likely to incur direct loss and/or expense due to deferment or as the result of one of the clause 4.24 relevant matters, they must make written application to the architect/contract administrator as soon as it becomes apparent that the regular progress of the works has been, or is likely to be, materially affected. In such circumstances contractors should obviously give the architect/contract administrator as much information as they can but, when required to do so, they have to submit to the architect/contract administrator such information as should enable him to form an opinion as to whether or not direct loss and/or expense has been incurred as a result of the matters cited in the application. Also, upon request, contractors have to submit to the architect/contract administrator or to the quantity surveyor such details as are reasonably necessary to enable them to ascertain the amount of loss and/or expense incurred or being incurred. Any amount so ascertained must be included in the next interim certificate and is not subject to a deduction for retention (see Chapter 28).

It must always be borne in mind that there can be an entitlement to an extension of time without an entitlement to reimbursement of loss and/or expense. Thus, although contractors may be granted extensions of time as a result of a delay caused by one or more of the clause 2.29 relevant events, they are not automatically entitled to reimbursement of direct loss and/or expense unless they can show that the cause of the delay was also a clause 4.24 relevant matter and that it actually (as distinct from theoretically) caused them to suffer direct loss and/or expense which could not be recovered under any other provision in the contract – for example through the valuation of a variation. Conversely, there can be an entitlement to reimbursement of loss and/or expense without an entitlement to an extension of time. This can occur when the relevant event is not likely to cause delay that affects the completion date but, as a clause 4.24 relevant matter, does cause the contractor to suffer loss and/or expense – for example delay to a non-critical element of the works, which causes additional expense due to loss of output.

Any direct loss and/or expense which in the opinion of the architect/contract administrator is incurred by the contractor as a result of compliance with the contractual provisions concerning any fossils, antiquities and other objects of interest or value found on the site, is reimbursable pursuant to clause 3.24. Whilst the clause 3.24 provisions relating to the ascertainment and reimbursement of direct loss and/ or expense are very similar to those in clause 4.23, they do not, unlike clause 4.23, make ascertainment and reimbursement conditional upon the contractor making written application to the architect/contract administrator.

Liquidated damages

It is a basic principle of English law that if one party to an agreement breaks a term of that agreement causing the other party to suffer financial loss, that second party can claim compensation or 'damages'. However, if it were left to aggrieved employers to claim and prove specific losses on each and every occasion that a contractor failed to meet a completion date, it would be a total waste of time, cost and effort in detailed litigation. It is therefore normal in building contracts to provide for employers to recover 'liquidated' damages, a set sum, from contractors if they fail to complete the works or section by the relevant completion date. However, to avoid such damages being set aside by the courts as a 'penalty', an unfairly onerous term in the contract, they must be 'ascertained', that is they must be a genuine pre-estimate of the likely loss that will be suffered by the employer if there is a delay to the completion date of the works or respective section. The pre-estimate of loss does not have to be totally accurate, but to be acceptable if queried in the courts it should be the result of making what is, in all the circumstances, a realistic attempt at pre-judging an employer's losses likely to arise from any delay between the contractual completion date and the actual date of practical completion.

The SBC makes provision for the recovery by the employer of liquidated damages in clause 2.32. It also requires the insertion of an amount and period for those damages in the contract particulars against clause 2.32. It is important that the entry in the contract particulars is carefully considered. As stated above, the amount of liquidated damages must be a genuine pre-estimate of loss and applies either to the whole of the works or a section of the works.

The operation for recovery of liquidated damages under the SBC is summarised below:

- If contractors fail to complete the works or section by the completion date the architect/contract administrator is required to issue a certificate to that effect.

- Provided that such a certificate has been issued employers may, if they so wish, inform the contractor in writing that they may withhold, deduct or require payment of liquidated damages. If the employer chooses to so inform the contractor, this must be done not later than five days before the final date for payment of the debt due under the final certificate.

- The employer may then advise the contractor in writing that the employer is to be paid liquidated damages at the rate stated in the contract particulars, or at a lesser rate if the employer so chooses, for the period between the completion date and the date of practical completion. Alternatively the employer may advise the contractor in writing that such liquidated damages will be deducted from monies due to the contractor.

- If, after the payment or withholding of liquidated damages, the architect/contract administrator fixes a later completion date under the procedures referred to above, the employer must pay or repay to the contractor any amounts recovered, allowed or paid in respect of the period between the original completion date and the later completion date.

Disputes and dispute resolution

As a consequence of building projects in general being complicated, unique ventures erected largely in the open, on ground the condition of which is never fully predictable and in weather conditions that are even less so, it is not uncommon for disputes to arise during the course of building operations. There are many reasons why disputes occur, but in the main they are caused by the failure of one or more members of the building team:

- to do their work correctly, efficiently and in a timely manner
- to express themselves clearly, or
- to understand the full implications of instructions given or received.

Most disputes are of a minor nature and are settled quickly, fairly and amicably by the building team. From time to time, however, more serious issues come into dispute. When this happens, the building team should make every effort to reach a fair settlement by negotiation. If this fails it becomes necessary to use one or more of the dispute resolution mechanisms available – mediation, adjudication, arbitration and litigation. Prior to adjudication under the Housing Grants, Construction and Regeneration Act 1996 (the Construction Act), arbitration was the most common method of resolving a dispute arising from a construction contract. However, since the Construction Act came into force on 1 May 1998 adjudication has become the most frequently used method of resolving a dispute.

Mediation

Mediation is a private, informal process in which one or more neutral parties assist the disputants in their efforts towards settlement. As with negotiation, the resolution lies ultimately with those in dispute – the mediator acts purely as a facilitator and cannot impose his decision on the parties.

Adjudication

A statutory process

Until statutory adjudication was imposed on 1 May 1998, the chief weakness of adjudication in the construction industry was that when push came to shove it was largely unenforceable. However, in the light of issues of cost and delay that were associated with arbitration and litigation (considered below), Sir Michael Latham's report *Constructing the Team* resulted in the Construction Act and Scheme in order to give swift, albeit temporary, justice by making the process of adjudication compulsory in most contracts. This is now the single most common mechanism by which construction disputes are resolved. Most construction contracts now contain adjudication provisions compliant with the Construction Act. If a contract were to contain provisions not compliant with the Construction Act, the Act itself would impose statutory adjudication giving a party to the construction contract the unilateral right to refer any dispute arising under the contract to adjudication.

Adjudication is now a binding process, available to the parties at any time, in which the party referring the dispute will receive a decision within 28 days of the referral irrespective of the complexity of the issues in dispute or the amount of money at stake. The 28 day period can be extended by 14 days if the adjudicator obtains the express approval of the referring party, but it can only be further extended if both parties so agree. The decision is binding and enforceable until the dispute is finally determined by arbitration or legal proceedings or by agreement. The parties may, however, agree to accept the decision of the adjudicator as finally determining the dispute. Even if the adjudicator makes a mistake in his decision, perhaps as a result of the sheer speed of the process, the courts will readily uphold and enforce it, leaving the disaffected party to raise separate proceedings in arbitration or litigation.

The SBC provisions

SBC Article 7 provides, for any dispute or difference arising under the contract, that either party may refer it to adjudication in accordance with clause 9.2. Clause 9.2 dispenses with the vast array of contractual rules that pervade its predecessor, JCT 98, by incorporating Part I Scheme for Construction Contracts (England and Wales)

Regulations 1998 SI/1998/649 (the Scheme), which is subordinate legislation to the Construction Act. Clause 9.2 simply requires that:

- for the purpose of the Scheme the 'adjudicator' is the person (if any) stated in the contract particulars but, in the absence of a name, the nominating body is identified
- the adjudicator deciding the dispute or difference must, where practicable, have the appropriate expertise and experience in the specialist area or discipline relevant to the issue in dispute
- in instances where tests are required for purported non-compliant aspects of the works, if the adjudicator is not appropriately qualified and experienced he must appoint an independent expert to advise in a written report as to the subject matter of the tests and dispute.

Notice of intention to seek adjudication

Under the Scheme:

- Any party to the construction contract may give written notice of his intension to refer a dispute (or difference as SBC clause 9.2 makes the distinction) arising under the contract to adjudication.
- The notice of adjudication must contain
  - — a description as to the parties involved in the dispute and the nature of the dispute
  - — details of where and when the dispute arose
  - — the redress sought from the responding party and
  - — specified contact details for each party for the purpose of giving notices.
- In the case of SBC clause 9.2, the person appointed as adjudicator or the nominating body (the professional body empowered to select the adjudicator) is in the contract particulars.
- Any person requested to act as an adjudicator must act in a personal capacity and not be an employee of any of the parties and, in order that the adjudicator is seen to be impartial, must declare any interest, whether it is financial or otherwise, in any matter relating to the dispute or difference.
- When a person has agreed to act as an adjudicator, the referring party must refer the dispute in writing to the adjudicator in seven days and must send copies of the documents to the responding party.
- The adjudicator may, with consent of all the parties, act on more than one dispute under the same contract, at the same time. Also, if all the parties agree, the adjudicator may act on one or more related disputes under different contracts.

- The parties to the dispute(s) may agree to extend the period within which the adjudicator may reach his decision in relation to all or any of disputes, whether connected or not.
- An adjudicator may resign his position at any time should the dispute be the same or substantially the same as a dispute previously referred to adjudication. Should an adjudicator resign his position for this reason, the referring party may start again promptly with a new notice and, if requested and insofar as reasonably practicable, supply the new adjudicator with copies of documents which were made available to the previous adjudicator.
- Should a party to the dispute object to the appointment of a particular person as adjudicator that objection does not invalidate either his appointment or any decision reached under his appointment.
- However the parties to the dispute may at any time jointly and unanimously revoke the appointment of the adjudicator. Should the parties agree to revoke the appointment of the adjudicator, they will be both jointly and severally liable for the payment of reasonable fees and expenses incurred by him before revocation, unless the revocation of his appointment is due to the adjudicator's default or misconduct.
- If the adjudicator's appointment is revoked or he resigns, the issues in dispute must be settled by agreement between the parties or referred under the Scheme to a new adjudicator.

Powers of the adjudicator

Under the Scheme:

- The adjudicator must act impartially in accordance with any relevant terms of the contract and in accordance with the applicable law in relation to the contract; and must do so without incurring unnecessary expense.
- The adjudicator may take the initiative in ascertaining the facts and law necessary for determining the dispute and must decide an appropriate procedure to be followed in the adjudication.
- The adjudicator may:
  - — request any party to the contract to supply him with such documents that he may reasonably require to supplement the referral notice or other documents that a party intends to rely upon
  - — decide upon the language to be used in the adjudication and whether translation of any document is required
  - — meet and question any of the parties or their representatives

- make any such site visits as he feels are appropriate, subject to any necessary third party consent
- undertake any necessary tests or experiments, subject to obtaining relevant consents
- obtain and consider such representations or submissions as he requires and, providing he has notified the parties of his intention, appoint experts, assessors or legal advisors
- give directions to be complied with as to the timetable for any deadlines or limits as to the length of written documents or oral representations.

- The parties must comply with any request or direction of the adjudicator in relation to the adjudication.

- Should a party fail to comply with the direction, request or timetable of the adjudicator, without sufficient cause, the adjudicator may:
  - continue the adjudication in absence of the third party or the document/written statement requested
  - draw such inferences from that failure to comply as circumstances may, in the adjudicator's opinion, be justified and
  - make a decision on the basis of the information before him, attaching such weight as he thinks fit to any evidence submitted to him outside any period he may have requested or directed.

- Unless the parties agree to the contrary in advance, any party, may be assisted or represented by such advisors or representatives as the adjudicator considers appropriate. However, unless the adjudicator gives directions to the contrary, when the adjudicator is hearing oral evidence or representations, a party to the dispute may not be represented by more than one such advisor or assistant.

- The adjudicator must consider any relevant information submitted to him by the parties and make available to the parties any information taken into account when reaching his decision.

- Except to the extent that it is necessary for the purposes of or in connection with the adjudication, the adjudicator and any party to the dispute must not disclose to any other person any information in connection with the adjudication where the party supplying it has indicated that it is to be treated as confidential.

- The adjudicator must reach his decision not later than:
  - 28 days from the notice referring the dispute to adjudication, or
  - 42 days from the notice referring the dispute to adjudication if the referring party consents, or
  - any such period exceeding 28 days from the notice referring the dispute to adjudication that both parties agree.

- Should the adjudicator fail, for any reason, to reach his decision within the aforementioned time periods, any party may serve fresh notice of their intention to refer the dispute to adjudication and, if requested by the new adjudicator, provide him with all documents made available to the previous adjudicator.
- As soon as practicable following the adjudicator's decision, he is required to deliver a copy of that decision to each of the parties in dispute.

Adjudicator's decision

- The adjudicator must decide the matters in dispute and may take into account any other matters that the parties agree should be within the scope of the adjudication or which are matters under the contract which he considers are necessarily connected with the dispute.
- The adjudicator may:
  - — open up, revise and review any decision made by any person referred to in the contract unless the contract states that any such decision or certificate is final and conclusive
  - — decide that any of the parties to the dispute is liable to make a payment under the contract and decide when such payment is due and the final date for payment
  - — having regard to the terms under the contract relating to the payment of interest, decide upon the circumstances in which, the rates at which, and the period for which compound or simple interest is to be paid.
- In absence of any directions as to the time of performance relating to the adjudicator's decision, the parties are required to comply with any decision of the adjudicator immediately upon delivery of his decision.
- The adjudicator is required to provide reasons for his decision if requested by one of the parties.

Effects of the decision

- The adjudicator may require, if he thinks it appropriate, the parties to comply peremptorily with his decision.
- The decision of the adjudicator is binding on the parties until the dispute is finally determined by legal proceedings, by arbitration or by agreement between the parties.
- The decision of the adjudicator is binding (see above) and the parties are required to comply with the decision. Should either party not comply, the other

party is entitled to commence an action for summary judgment to secure such compliance.

- The parties are jointly and severally liable for the payment of the adjudicator's fee and reasonable expenses. The adjudicator may direct how this is to be apportioned between the parties. If he does not do so the cost is borne equally by the parties.
- Unless given express powers, the adjudicator has no power to award the 'winning' party its costs; as such, each party will bear its own costs.
- The adjudicator is not liable for anything he does or omits to do in the discharge of his functions as adjudicator unless he acts in bad faith.

Arbitration

An alternative to litigation

Arbitration is a process, subject to statutory controls, whereby a private tribunal of the parties' choosing formally determines a dispute. It became a preferred alternative to litigation in the second half of the 20th century because it allowed the parties privacy and an element of control. Arbitration was also, to begin with at least, a more expeditious option, although the process later became beset with the same problems as with litigation, namely delaying tactics by one of the parties.

Many people in the construction industry still take the view that it is essential to have their disputes heard by someone with a working knowledge of building contracts and the practices that are prevalent in the industry. They often lean towards arbitration as a means of achieving that end. (Although in doing so they fail to take account of the expertise available within the Technology and Construction Court (TCC), which is almost invariably the forum within the High Court structure that hears building contract matters – see below.) Arbitration, therefore, essentially represents a process that is available as an alternative to litigation with the purported advantages of flexibility, economy, expedition, privacy and freedom of choice.

SBC Article 8, where it applies, provides for either party to refer to arbitration any dispute or difference arising under the contract, either during the progress or after the completion or abandonment of the works or after the termination of the contractor's employment, except:

- any disputes or differences arising under or in respect of the Construction Industry Scheme or VAT, to the extent that legislation provides for some other method of resolving the dispute or difference
- disputes in connection with the enforcement of any decision of an adjudicator.

This provision is an alternative to Article 9 which, by default, provides for the settlement of disputes or differences by legal proceedings. If the arbitration alternative is chosen it is to be set down in the contract particulars. When arbitration applies, the relevant provisions are to be found in clauses 9.3 to 9.8 of the SBC.

Conduct of arbitration

The provisions of the Arbitration Act 1996 apply to any arbitration under the contract and any such arbitration is to be conducted in accordance with the JCT 2005 edition of the Construction Industry Model Arbitration Rules (CIMAR) current at the contract base date.

Notice of reference to arbitration

If either party requires a dispute or difference to be referred to arbitration then that party has to serve on the other party a written notice of arbitration identifying the dispute and requiring him to agree to the appointment of an arbitrator.

- If the parties fail to agree on an arbitrator within 14 days of the service of the notice of arbitration then either party may apply to the person or body named in the contract particulars for the appointment of an arbitrator.
- If the previously appointed arbitrator ceases to hold office, for any reason, the parties may agree upon an individual to replace him or either party may apply to the appointor for the appointment of a replacement.
- Where two or more related arbitral proceedings fall under separate arbitration agreements, the persons who are to appoint the arbitrator may appoint the same arbitrator for all related proceedings.
- After an arbitrator has been appointed either party may give notice to the other and to the arbitrator referring any other dispute under the contract to be decided in the arbitral proceedings.

Powers of arbitrator

The arbitrator has to determine all matters in dispute that are submitted to him and in doing so he has the power to:

- rectify the contract so that it accurately reflects the true agreement made by the parties
- require such measurements and valuations as he considers necessary to determine the rights of the parties

- ascertain and award any sum that ought to have been included in any certificate
- open up, review and revise any certificate, opinion, decision, requirement or notice issued under the contract.

Effect of an award

- Subject to an appeal on any question of law arising out of an arbitral award, the award of the arbitrator is final and binding on the parties.
- Either party is at liberty to apply to the courts to determine a question of law arising in the course of a reference.
- Subject only to the right of either party to appeal to the courts, any question of law arising out of an award of the arbitrator is final and binding.

Litigation

The courts provide the setting for the traditional mode of dispute resolution – namely litigation. However, with the past developments of arbitration and then adjudication the number of disputes actually determined by the courts has declined compared with those settled by other means. Furthermore, very few of the proceedings that are begun actually result in a full trial and subsequent judgment. In fact, more than 90% of actions begun in the High Court are disposed of before reaching trial.

Most building disputes are dealt with by what were previously known as the Official Referees, now known as the Technology and Construction Court (TCC) judges who are, almost without exception, appointed from leading counsel and spend a great deal of their time hearing disputes about building contracts. It is therefore worth noting that the TCC judges have built up an expertise over the years to rival that of the most eminent professional sitting as an arbitrator and, unlike the set up with arbitration, litigants do not have to pay the judge and his support staff or hire the court room.

In his 1996 report *Access to Justice*, Lord Woolf expressed the opinion that the then current system of litigation was too expensive, too slow, too fragmented, too adversarial, too uncertain and incomprehensible to most litigants. As a result of his review, the Civil Procedure Rules were implemented on 26 April 1999 (replacing the former County Court Rules and Rules of the Supreme Court) and apply to all actions begun from that date. It is worth just a mention here that a key facet of these rules was that they were designed to help ensure that all cases are dealt with justly and in ways that are proportionate to the amount of money involved, the importance of the case, the complexity of the issue and the parties' financial position.

There are now three 'tracks' on which litigation runs:

- the small claims track for disputes of less than £5,000
- the fast track for disputes between £5,000 and £15,000
- the multi-track for disputes over £15,000.

SBC Article 9 provides that, subject to Article 8, the English courts have jurisdiction over any dispute or difference between the parties which arises out of or in connection with the contract. Legal proceedings, subject always to a party's right under the Construction Act to refer a dispute to adjudication at any point in the contract, is the default position for finally determining disputes or differences under the contract. The contract particulars relating to Article 8 state that if disputes are to be determined by arbitration and not by legal proceedings, it must be positively stated that Article 8 and clauses 9.3 to 9.8 apply.

Notification of Revision to

Completion Date

JCT 98

Issued by: Ivor Barch Associates
address: Prospects Drive, Fairbridge

Employer: Cosmeston Preparatory School
address: Fairbridge

Job reference: IBA/94/20

Contractor: L&M Construction Ltd
address: Ferry Road, Fairbridge

Notification no: 6

Issue date: 6 January 2000

Works: New School Library
situated at: Park Street, Fairbridge

Contract dated: 14 December 1998

Under the terms of the above-mentioned Contract,

I/we give notice that the Completion Date for

*Delete as appropriate

* the Works
~~* Section no. ________ of the Works~~

previously fixed as

17 December ~~20~~ 1999

* is hereby fixed later than that previously fixed,
~~* is hereby fixed earlier than that previously fixed,~~
~~* is hereby confirmed,~~

and is now

21 January 20 00

* This revision has taken into account the following Relevant Events:
Exceptionally adverse weather conditions (cl. 25.4.2)
Compliance with the Architect's instructions nos. 46 and 47 (cl. 25.4.5.1)

~~* This revision has taken into account the omission of work required by the following Instructions:~~

~~* This revision is made by reason of my/our review.~~

To be signed by or for the issuer named above

Signed *Ivor Barch*

Distribution

☐ Contractor	☐ Quantity Surveyor	☐ Clerk of Works	☐
☐ Employer	☐ Structural Engineer	☐ Planning Supervisor	☐
☐ Nominated Sub-Contractors	☐ M&E Consultant	☐	☐ File

F808 for JCT 98

Example 30.1 Revision to completion date.

Chapter 31
Insolvency

Introduction

In simple terms, insolvency means being unable to pay debts as and when they become due. It is a liquidity or cash flow problem which can affect even wealthy companies in difficult trading conditions. This chapter deals with corporate insolvency under building contracts, rather than bankruptcy, which is individual insolvency applying only to sole traders and partnerships.

The most common procedures of insolvency for companies are liquidation, administrative receivership and administration. The Enterprise Act 2002, since its coming into force on 15 September 2003, introduced changes to advocate administration as the method in which nearly all insolvent corporate businesses should be rescued or resold.

Unlike many other events which occur during the course of a contract and can be planned for, insolvency is one over which the building team has no control and thus the procedure consequent upon insolvency can be actioned only after the event. While insolvency can overtake either party to a contract, generally speaking it happens more often to the contractor than to the employer. This chapter therefore concentrates on insolvency of contractors, but the insolvency of employers is considered briefly at the end.

Before considering the procedure which is laid down within the SBC, it is important to have an understanding of insolvency generally, in order to put that procedure in its proper context. It is also important to realise that the contractual provisions must be interpreted within the complications of general insolvency law and the parties may therefore need to seek expert advice from an Insolvency Practitioner.

Insolvency practitioners

Insolvency practitioners are licensed by their professional bodies (e.g. The Law Society or The Consultative Committee of Accountancy Bodies) and, since the

Insolvency Act 1986, any person acting as a liquidator, provisional liquidator, administrative receiver, administrator or nominee/supervisor of a voluntary arrangement must be an insolvency practitioner.

Liquidation

The liquidation or winding up of a company is a legal process that brings its existence to an end in order that its assets can be realised for the benefit of its creditors and members. It is the corporate equivalent of dying. There are two types of liquidation in respect of companies: voluntary and compulsory.

Voluntary liquidation arises in two situations:

- **A members' voluntary liquidation** occurs when the members or shareholders of the company at an extraordinary general meeting resolve to wind up the company, which is still warranted by the directors to be solvent. More often than not, a members' liquidation is not an unforeseen event. It is often used when restructuring company assets within a group or parent holding company.
- **A creditors' voluntary liquidation** can, however, arise if the company is considered to be insolvent, as a creditors' meeting will follow the shareholders' meeting so that a liquidator can be appointed.

Compulsory liquidation arises when one or more creditors petition the court for the company to be wound up or placed in liquidation. In this case, if the court considers that the company is unable to meet its debts, it will order the winding up. Then a provisional liquidator, usually the official receiver, is appointed by the court pending the appointment of an official liquidator by the creditors at a meeting. It is possible for the official receiver to be appointed to the post, but it is more common for an independent insolvency practitioner to be appointed, except in 'no asset' cases when the official receiver continues to act because there are no assets to pay a private practitioner.

Once a liquidator is appointed and in office, he will:

- identify assets of the company and have them valued
- deal with proper claims from third parties that assets in the possession of the company that have not been paid for are subject to retention of title (a 'Romalpa clause') or creditors' claims that bank accounts or property were held in trust for them
- invite creditors to submit details of their claims as 'proof of debt'
- sell company property
- distribute the net proceeds of the sale of property in order of priority – 'the statutory scheme'

- investigate the conduct of directors and, depending on their individual and collective conduct, file a report on their fitness to act as a director in the future
- pursue property or money misappropriated during the run-up to the company's liquidation.

Receivership

Receivership is a remedy afforded to creditors who have obtained a charge over the whole or part of a company's property and is governed by the Insolvency Act 1986. The process of receivership can be ordinary or administrative and is a process whereby an insolvency practitioner is appointed under specific instructions, which need not necessarily end in liquidation. A receiver will usually achieve a better realisation of the assets than a liquidator, being able to trade the asset(s) and sell the goodwill, thus obtaining the best value possible consistent with the time and cost involved in achieving the realisation.

In ordinary receivership, a receiver is appointed by a debenture holder, often a bank, under the express provisions of the debenture (see below). His job is to realise the asset(s) over which he has control to satisfy the debt due to the appointor, after which he withdraws and any surplus proceeds are returned to the company. In practice, receivership often leaves the company as a mere shell without assets and more often than not leads to liquidation if the residue of the company is incapable of trading.

In administrative receivership, the administrative receiver is appointed by the holder of a debenture secured by a floating charge (typically a bank or other lending institution). The appointment has immediate effect, there being no waiting period pending a court decision or meeting of creditors, and imposes an immediate stop on payments by the company to its creditors. The administrative receiver's powers are derived from the debenture and the Insolvency Act 1986 and involve taking control of the whole company (not just a particular asset). The principal duty of the administrative receiver is to realise the assets charged by the debenture for the benefit of the secured creditor, but only after the preferential creditors (either those with a fixed charge or statutory entitlements who have legal priority) have been satisfied. Any surplus funds realised will be returned to the directors of the company, or to any liquidator appointed, for distribution to the unsecured creditors and shareholders of the company.

Administration

Administration is a procedure instigated by the Insolvency Act 1986 and was radically reformed when the Enterprise Act 2002 came into force on 15 September 2003. Administration is a court based procedure, whereby an administrator is appointed by the court following a petition to the court, which is almost always brought by the directors of the company.

There are two prerequisites for the granting of an administration order – first, that the company is, or is likely to become, insolvent, and second, that the creditors' interests will be served thereby. Although the agreement of any debenture holders is not required, in practice they have to be persuaded, otherwise they can individually enforce their own security and appoint an administrative receiver, forestalling the court appointment. The purpose of the court appointment is to protect the company from its creditors so that a rescue package can be put in place – a moratorium being placed on the payment of the creditors. The role of the administrator is to keep the company functioning as a going concern and, hopefully, to trade the company out of its financial difficulties within a given timescale. If unsuccessful, the company may eventually have to go into liquidation.

An administrator hopes to preserve the assets of the company while satisfying the creditors, but where realisation has to take place, he will usually achieve a better realisation than a liquidator or receiver, as he will often restructure the business prior to disposal, obtaining the best value possible, consistent with the time and cost involved in achieving the realisation.

After the Enterprise Act 2002

Administration had not been as successful as the government originally hoped in setting up the scheme under the Insolvency Act 1986 – it was often frustrated by debenture holders objecting to the appointment of an administrator, and was expensive to operate due to the high setting-up and running costs. It is for this reason that Parliament intervened with the Enterprise Act 2002.

Aims of the regime

The new administration regime brought in under the Enterprise Act 2002 incorporates a Schedule B1 into the Insolvency Act 1986. These provisions now specifically require an administrator, who is an officer of the court, to perform his function with the following objectives:

> '*(a) rescuing the company as a going concern, or*
> *(b) achieving a better result for the company's creditors as a whole than would be likely if the company were wound up (without first being in administration), or*
> *(c) realising property in order to make a distribution to one or more secured or preferential creditors. (Insolvency Act 1986, Schedule B1, 3(1))*'

In respect of these objectives, the administrator must rescue a company as a going concern unless he perceives that it is not reasonably practicable to do so or that winding up the company would achieve a better result for the company's creditors as a whole. An administrator should only entertain the last of these three objectives if it is not reasonably practicable to realise either of the first two and he does not unecessarily harm the interests of creditors as a whole.

Appointment of an administrator

Under the new regime, there are three methods of appointing an administrator:

- first, a court can appoint an administrator – but only on the application of:
    - the directors of the company
    - one or more creditors
    - an official of a magistrates court for a fine imposed on the company
- second, by a qualifying floating charge holder without involving the court
- third, by the company or its directors without involving the court.

In respect of the second point, the administrator may only be appointed if the floating charge purports to empower the charge holder to appoint an administrator or an administrative receiver. The charge must also relate to the whole or substantially the whole of the property. In order for the appointment to take effect, the notice of appointment must be filed in court and the appointer must make a statutory declaration, which includes a statement that the floating charge relied upon was enforceable on the day of the appointment.

A company or its directors can appoint an administrator but may only do so upon giving five business days notice to any person who is or may be entitled to appoint an administrator or an administrative receiver (i.e. the holder of any debenture secured by a qualifying floating charge). The notice must be filed in court and is effective from such a time that the notice is received by the court. Together with the notice of intention to appoint an administrator, the application must be accompanied by a statutory declaration including a statement that the company is, or is likely to become, unable to pay its debts.

The administrator appointed has eight weeks in which to appraise the status of the company and send out proposals to creditors. Following distribution of his proposals, the administrator is required to hold a creditors' meeting to discuss them.

Voluntary arrangements and compositions

When a company runs into cash flow problems, there is usually a period before insolvency is reached when, by planning cash flow on the basis of extended credit, the company can work itself out of its difficulties and back into normal trading. The formalisation of these extended credit facilities is called 'an arrangement for a composition of debts' or 'a scheme of arrangement'. The Companies Act 1985 (amended by the Companies Act 1989) and the Insolvency Act 1986 give statutory backing to such rescue arrangements after a strict procedure has been followed.

The procedure is initiated by the company proposing the arrangement under the supervision of an insolvency practitioner, called a 'nominee'. The nominee must prepare a report for the court and will decide whether a meeting of the company and creditors should be called (and the arrangements therefor). In order to be implemented, the proposal must have the support of 75% in value of the creditors. Once approved the nominee becomes the 'supervisor' of the arrangement.

It should be noted that the Acts do not give an immediate moratorium on creditor demands and so the process can be undermined by any individual creditor pursuing its individual rights before the arrangement is agreed. Further, such arrangements are effective only against creditors who have received notice of the proposed arrangement, and in any event a creditor having a debenture secured by a floating charge maintains the right to appoint an administrative receiver at any time.

Debentures

A debenture is a bond, evidencing a debt, which contains conditions as to the interest payable and its settlement, discharge or satisfaction. A debenture is generally secured by a fixed or floating charge on a company's assets registered at Companies House. A fixed charge identifies a particular asset which can be liquidated to pay the debt; a floating charge is a charge on the general assets of the company. When the asset subject to a fixed charge is a building, the debenture is commonly arranged as a mortgage.

Fixed and floating charges rank as secured, or preferential, creditors in a winding up. Fixed charges rank first, followed by statutory responsibilities, such as unpaid wages, PAYE and VAT, and then floating charges. The unsecured creditors such as suppliers are, of course, bottom of the list and often receive nothing at all.

Procedure upon the insolvency of the contractor

It is important to understand that the procedure is carefully codified within the SBC and the detailed wording of clause 8.5 of the contract should be studied in addition to the general advice given in this chapter.

Ordinarily, and without express words to the contrary, becoming insolvent is in itself not a breach of contract. However, clauses 8.1 and 8.5, by defining 'insolvency', provide express words allowing the employer to conclude that there is an anticipatory breach of contract by the contractor, i.e. the employer anticipates that the contractor may not be able to complete its obligations and may, therefore, terminate the contractor's employment.

Clause 8.1 draws a distinction between those insolvency procedures which are orientated towards business rescue and those which lead to the inevitable demise of the company, whilst also recognising that a company may enter into an arrangement 'as a solvent company for the purpose of amalgamation or reconstruction'. It is in the employer's interest that termination of the contractor's employment in the event of business rescue being pursued is only an option for the employer and that the contractor's employment is not automatically terminated – in the hope that work on site will not stop, that a rescue bid will preserve the company and that this will lead to the most economic completion of the works. In the case of the ultimate demise of the company, the automatic determination of the contractor's employment remains.

The business rescue-orientated insolvency procedures include administrative receivership, administration and voluntary arrangements or compositions. The insolvency procedures leading to the inevitable demise of the company include the appointment of a provisional liquidator, compulsory liquidation and a creditors' voluntary liquidation.

The SBC provisions

Section 8 provides for the termination of the contractor's employment in the event of default (8.4) and corruption (8.6) as well as in the event of insolvency (8.5). The procedure to be followed by the employer and the contractor is set out in clauses 8.4, 8.5, 8.7 and 8.8.

Clause 8.1 defines the term insolvency under the contract and, in doing so, makes the distinction between those proceedings that are ordinarily expected when a contractor becomes insolvent (enters into arrangements for the satisfaction of debts) and those that apply in the case of company restructuring. This distinction is important. If a contractor makes an arrangement for the purpose of company restructuring, it is not 'insolvency' as defined by the SBC and the employer cannot give notice terminating the employment.

However, if the contractor is, for the purposes of the contract, insolvent under clause 8.1, then under clause 8.5.1, while the contractor's employment is not automatically determined, the employer may at any time, by notice, terminate it. The practicalities of how the employer is expected to establish that the contractor is insolvent according to the definition in clause 8.1 is covered in clause 8.5.2. This requires that contractors shall immediately inform the employer in writing if they make any proposal, give notice of any meeting or become the subject of any proceedings or appointment as covered by the definition of insolvency in clause 8.1.

Clause 8.5.3 relieves employers from their obligations to make further payments to the contractor and suspends the contractor's obligations to carry out and complete the works. However, the clause also requires employers to ensure that the site, the works and the materials on site are adequately protected.

Clause 8.7 deals with the consequences of terminating the contractor's employment. The clause covers the completion of the works, the contractor's designed portion (where applicable), the making good of defects, the use of temporary buildings and plant, assignment of the benefit of any agreement for the supply of materials or goods and/or for the execution of any work and the computation and settlement of the final account.

Clause 8.8 deals with the procedure in the event that the employer decides not to complete the works (see below under the heading *Final accounts*).

Early warning signs and precautionary measures

The design team will usually be aware of the contractor's financial difficulties before any formal steps are taken in recognition of insolvency. There may be rumours circulating around the site. There may be a rapid depletion of the contractor's resources, such as materials, temporary buildings or the workforce. However, although contractors may be, or appear to be, in financial difficulties, they need not necessarily be insolvent. In these circumstances the matters should be kept strictly confidential, not only to avoid the possibility of a defamation action but also to prevent the spread of such information, which could quickly drive a contractor into more serious difficulties or into liquidation.

The first moves . . .

Clause 8.5.1 requires contractors to give written notice to the employer where they:

- enter into an arrangement, compromise or composition with creditors, or
- make a determination (without a declaration of solvency) to be wound up, or
- have a winding up order made against them, or
- have an administrator or an administrative receiver appointed – the business rescue scenario.

An employer should take immediate steps to discuss the situation with the contractor and any appointed insolvency practitioner. Any formal notice given by employers under clause 8.5.1 should clearly state that they regard the employment of the contractor terminated. This is to avoid any suggestion that, by carrying on discussions, the employer has waived the right to terminate the contractor's employment under the contract. However, this may be a moot point as clause 8.5.1 states that if the contractor is insolvent, the employer may at any time issue a notice.

It is also important to note that it is only the 'employment' of the contractor which is determined, not the contract itself. The contract continues in order to protect the rights of the parties and govern the procedure to be followed.

The principal actions which the design team should take, on behalf of the employer, within the first few days following the insolvency of a contractor are as follows:

- secure the site as far as possible in a cost-effective way to prevent vandalism and unauthorised removal of materials by sub-contractors and suppliers

- list all unfixed materials (whether or not 'site materials'), with a note of those paid for by the employer
- establish the position regarding unpaid certificates, retention, performance bonds and other contractual entitlements
- establish with professional advisers the objectives and determine the employer's preferred route forward
- make arrangements, if appropriate, to allow the contractor, sub-contractors and suppliers access to the site during working hours to continue with the works
- prepare a programme and an estimate of cost for the completion of the works.

Safeguarding the site, the works and materials on site

Ideally as soon as the contractor's employment has been terminated, the site should be closed and no materials or plant should be removed, despite the various demands of suppliers and sub-contractors for the return of materials. An understanding of the position regarding ownership of materials is therefore important.

It is a well-established rule that ownership of all materials and fittings, once incorporated in or affixed to a building, passes to the owner of the freehold. Although the employer may not be the freeholder, in practice this should make little difference to this rule. Ownership in unfixed materials and goods, however, can be more complex and there is no easy method of determining title. The relevant clauses in the SBC are 4.17 (off site materials and goods) and 4.6 to 4.15 (certificates and payments).

Clause 2.24 provides for the ownership of unfixed materials and goods to pass to the employer, where their value, in accordance with clause 4.16, has been both included in a certificate and paid for by the employer.

The architect/contract administrator has a duty to certify the total value of the materials and goods delivered to or adjacent to the works, provided that they have been properly and not prematurely delivered and are adequately protected (clause 4.16.2), but he may only certify materials or goods before delivery if they are the 'listed items' (clause 4.17). One might list items such as steelwork, mechanical and electrical equipment or any materials and/or goods which are in great demand or have long manufacturing lead-in times. These items usually have to be ordered and manufactured in time slots which do not run conveniently in parallel with the project but must, obviously, be ready for use when required.

Clause 3.9.2.1.1 provides that, where the value of a sub-contractor's materials has been included in a certificate and this has been paid, such materials or goods shall become the property of the employer, and the sub-contractor shall not deny this. Provision is also made within the SBC for the ownership of materials and goods to pass to the contractor where the contractor has paid a sub-contractor in advance of the architect/contract administrator certifying such payment (clause 3.9.2.1.2 refers).

In practice, it is probably safest to assume that all materials on site are in the ownership of the employer and to allow nobody on to the site under any pretext whatsoever. It will then be up to the liquidator or administrator or to a sub-contractor or supplier to prove that ownership in particular materials has not become vested in the employer. There is, of course, always the risk that, if the title to materials and goods has not passed to the employer, the supplier or sub-contractor may seek an injunction for their release or damages for retention or conversion. If there is the possibility of such a dispute, legal advice should be sought.

If a clerk of works is employed, he may be able to act as a watchman while he is on site. However, it would be prudent for the employer to engage a separate watchman to cover periods when the clerk of works is away from the site. It would also be prudent for the quantity surveyor to make a list of the unfixed materials on the site – he will need this in any event in determining what is necessary to complete the works. The project manager or the head of the design team should immediately try to arrange with the employer for all valuable materials which can be moved to be secured under lock and key and for the locks to be changed where materials are already secured.

'Listed items' off site

Clause 2.25 provides that, where the value of any 'listed items' intended for the works and stored off site has been certified in accordance with clause 4.17 and paid for by the employer, such 'listed items' shall become the property of the employer. However, the question of title in respect of materials and goods stored off site can be complicated. For example, the contractor would have no title to the materials and goods where these are the subject of a retention of title clause.

The use of such a clause was first tested in the courts in the case of *Aluminium Industrie Vaassen BV* v. *Romalpa Aluminium Ltd* (1976). In its most basic form, a 'Romalpa' clause will state that title to the materials and goods will be retained by the supplier unless and until they have been paid for. The issue may be further complicated where the supplier is nevertheless divested of title to the goods by virtue of section 25 of the Sale of Goods Act 1979. This section provides, in effect, that the contractor can, by reselling the goods to a bona fide purchaser as a 'buyer in possession', pass on ownership, even though the contractor lacked good title at the time. This is a complicated area of the law and it would be wise to seek legal advice in the event that a dispute over title becomes likely.

Completion of the works

It will be necessary to arrange a meeting with all concerned to consider the best method of completing the contract. The employer, professional advisers and all consultants should be at this meeting. When a liquidator, administrator or administrative receiver has been appointed he should be kept informed of decisions which have been made. Similarly, if a bond is coupled with the contract, the bond holder should be kept informed. It would be wise, in the light of court decisions, for the employer to check the terms of the bond immediately on becoming aware that the contractor is in financial difficulties. All requirements under the bond, for example as to notice, should be complied with.

Assuming that the contractor's employment has been terminated, the most obvious way of completing the contract is employing another contractor, if possible on terms similar to those of the first one.

Undertaking to complete the works

Following termination of the contractor's employment, the provisions of clause 8.7 become operative, allowing employers to complete the works as economically and expeditiously as possible – assuming that they wish to do so rather than decide not to under clause 8.8.

To facilitate the completion of the works, clause 8.7.1 provides for the employer to employ and pay other persons to carry out and complete the works. Such persons may use all temporary buildings, plant, tools, equipment and materials and purchase all additional materials necessary to complete the works and make good defects. However, these rights of the employer, when based upon a termination through insolvency, are dependent upon

- first, a valid right in law to use such buildings, plant, tools, etc., and
- second, the owners of the various items being bound by the contract.

In the first case, it is often argued that this clause is void against a liquidator who, as a matter of law, upon his appointment takes into his custody and control all the property of the company. In the second case, many items of plant are hired by the contractor or owned by a subsidiary company which is not a party to the contract. In these circumstances it will be necessary to enter into new agreements with the owners of such equipment.

All these factors will need to be taken into account in putting together the documentation and agreeing the arrangements with the completion contractor. Much will depend on the extent of the works to be completed:

- If the works have not started, it will be necessary to appoint a contractor and to fix a new price for the works. This can be done by negotiating, if possible, with the second lowest tenderer or, if that fails, by seeking new tenders, preferably from those contractors on the original tender list. In this instance there should be no need to amend the original tender documents other than to revise the dates for possession and completion.

- If the works have started but are not complete, it will be necessary to appoint a contractor and to fix a price for the completion works. This can be done either by negotiating with a single contractor or by seeking competitive tenders. In either instance the original tender documents can be used but they will need amending so as to reflect:
  - — the extent of the works properly completed by the original contractor prior to determination
  - — the materials, goods, plant and equipment available on site for use in the completion works
  - — the method for dealing with the making good of defects arising from the original contract (see later in this chapter)
  - — the new dates for possession and completion
  - — any appropriate revision to the amount of liquidated and ascertained damages.
- If the works are substantially complete or complete apart from the making good of defects, it will still be necessary to appoint a contractor for the completion works. In this situation the remaining work is likely to be of a jobbing nature, which is perhaps best done under a prime cost contract with a fixed fee or, if available, by the employer's own direct labour organisation or maintenance department. In either instance completely new tender documents will be required.

In any event the design team will need to carefully assess the work to be done, including that of sub-contractors, before making recommendations as to the most suitable procedure and contract for completion works. Employers should ensure that completion works are carried out in the most economical way if disputes with any bond holder or liquidator are to be avoided. They must also ensure that, for the purposes of CDM94, a principal contractor is always in post while any work is in progress on the site (see Chapter 1).

Completion documentation

One item that is essential in any scheme for completion of the work, except where the contractor's employment is reinstated, is a method for dealing with the making good of defects which are bound to arise from the original contract. There are three options available:

- a lump sum premium
- a provisional sum
- measured quantities for known defects, plus a lump sum premium or provisional sum for other items.

Lump sum premium

In certain cases, the most satisfactory completion document is one on identical terms to the original contract, with the addition of a fixed lump sum premium to cover the new contractor's costs in taking over the partly completed works, including price fluctuations, the responsibility for making good defects and a credit for any materials, goods, plant and equipment on site which belong to the employer. This lump sum premium is then paid to the contractor by instalments as the work proceeds. This method has the advantage of identifying the employer's total commitment at the start and avoids the need to account for the cost of remedial works and therefore saves professional fees. Unfortunately, this method cannot be used if the new contractor is not able to make a proper assessment of the risks involved – usually where the remedial works cannot be defined.

Provisional sum

Where the remedial works cannot be defined, a provisional sum has to be included to cover the cost of any such work which comes to light during the course of the completion contract. The SBC requires the architect/contract administrator to issue instructions in regard to the expenditure of provisional sums. Variations arising from such instructions fall to be valued according to the normal rules for the valuation of variations (see Chapter 27). This system has the disadvantage of an unknown commitment by the employer, but avoids the new contractor having to assess, and therefore price, the risk of carrying out the remedial works arising from the former contract.

Measured quantities for known defects, plus lump sum premium or provisional sum

When certain defects are known to involve extensive remedial work it is good practice to incorporate details of the necessary remedial work in the tender documents for the completion works. The cost of making good the remaining defects can then be incorporated either by way of a lump sum premium or a provisional sum. This is a 'halfway house' solution, which has the advantage of identifying a large element of the remedial works costs and reducing the element left at risk to either the contractor (in the case of a premium) or the employer (in the case of the provisional sum). It may well be an appropriate compromise in an uncertain situation, when time available to reach an agreed basis for completing the contract is limited.

Sub-contractors

Under SBC clause 3.9, it is a condition that if the contractor is sub-letting part of the works, then the sub-contract must contain provisions immediately terminating the sub-contractor's employment on termination of the contractor's employment. When this occurs there are bound to be a number of sub-contractors to deal with and they will fall into two groups:

- Where the sub-contract works have not been started, it might be prudent to invite them to enter into a similar sub-contract with the new contractor when the new contract is placed.
- Where the sub-contract works have been partially completed the quantity surveyor should negotiate a price for completing the work, taking care to ensure that only the work to be completed is measured, assessed and valued and making no allowance for the possibility that the sub-contractor may not have been paid for the work already done.

Clearly, it will be in the employer's interest for the sub-contract works to be carried out by the original sub-contractors, but it may be that they will not agree to complete unless they are paid, in full, for all the work carried out under their original sub-contract agreements. In such instances the design team will have to assess whether or not it would be more cost effective to have work completed by others.

Bonds

The contractor might have been required to provide a bond (through a bank or insurance company) for the due completion of the work and this is usually set at 10% of the contract sum. Subject to its terms, such a bond was traditionally thought to be available to be used when the employer was put to additional expense as a result of a contractor's liquidation. However, this is now by no means certain in view of findings by the courts that the insolvency of the contractor does not necessarily amount to a breach of contract, as there are procedures dealing with it within the contract. Until the uncertainty is removed, this is obviously an area where legal advice should be sought by or on behalf of the employer.

Although sureties may be liable for paying the employer such extra money as may be necessary to complete the job, they have no control over the way the contract is completed. However, any employer should complete the work without undue extravagance and it would be prudent to keep the surety informed as the work proceeds.

Final accounts

SBC clause 8.7 deals with the respective rights, duties and responsibilities of the employer and the contractor. While this could mean that the contractor would be entitled to payment, in practice this is seldom the case. The employer is entitled to the additional expenses of completing the works, and to damages for any loss suffered by reason of the termination of the contractor's employment. It appears that this would include any extra fees and expenses which are properly incurred, and the costs of delay. For example, the architect/contract administrator may well have to make additional visits to the site, and additional prints of drawings and contract documents may be required. The quantity surveyor will be involved in a great deal of extra measurement, as well as having to sort out complex final accounts.

Insolvent contractors are, by the provisions of clause 8.7, responsible for defraying these expenses properly incurred by the employer as well as the amount of any expense caused to the employer by termination of their employment. This is, though, subject to receiving a credit for what would have been paid had the contract not been terminated.

Two final accounts, therefore, have to be prepared:

- The first, or notional, final account will be the amount that the contract would have cost had the first contractor completed in the ordinary way. This should include variations ordered during both the original and the completion contract, all priced at the rates relevant to the original contract.
- The second final account will be a normal final account for the completion contract, but will include only those variations ordered during the completion contract. These will be priced by reference to the pricing structure of the completion contract, the rates of which are likely to be different from those in the original contract.

Care should be taken to ensure that if the completion contractor has to make good any defects arising from the original contract, the cost of such works, priced as a premium or treated as variations on the completion contract, does not appear as a variation in the first final account; this is because they would not have been so included if the first contractor had completed normally.

The agreement of the second final account should not create any difficulties because there is a contractor's organisation with which to work. But the first final account will almost certainly have been prepared by the quantity surveyor without the assistance of the first contractor, whose staff, following liquidation, will probably no longer be available. The liquidator should therefore be kept informed of the method used to prepare the account and should be sent a copy on completion for his information.

Two examples of a statement showing the financial position of the parties at completion are given at the end of this chapter. In Example 31.1, it has been assumed that the contract was completed without too much trouble. It has also been assumed that the extra cost of completing the contract was less than the monies outstanding to the contractor at the time of the liquidation and therefore there is finally a debt due from the employer to the liquidator (an unusual outcome but a possibility nevertheless). In Example 31.2, on the other hand, it has been assumed that, with the contract only one third complete, the extra cost of completing the work was considerably greater than the money outstanding at the time of the liquidation. Almost certainly, the employer suffers a loss; how much of this he can recover depends on the amount in the pound paid by the liquidator and any amount contributed by the bond holder, if applicable. In either example, one is seeking to compare that which the employer pays out with that which would have been paid out but for the contractor's demise.

Clause 8.8 covers the procedure in the event that employers decide not to complete the works. In such circumstances, they are required to notify the contractor within a period of six months and thereafter, within a reasonable time, to send a statement of account to the contractor.

The procedure upon the insolvency of the employer

Clause 8.10 of the SBC deals with the insolvency of the employer in a manner similar to the way that clause 8.5 deals with the insolvency of the contractor.

Clause 8.10.1 requires employers to inform the contractor in writing if they make any proposal, give notice of any meeting or become the subject of any proceedings or appointment relating to the contract definition of insolvency.

Clause 8.10.1 provides for contractors to give notice to the employer terminating their employment on the employer becoming, for the purpose of the conditions, insolvent. Notwithstanding the contractor's obligation to give notice to the employer, the contractor's obligations to carry out and complete the works or the design in a contractor's designed portion under clauses 2.1 and 2.2 are suspended.

Clause 8.12 covers the consequences arising from termination for any reason, including the employer's insolvency. In essence, the contractor is required to remove all his temporary buildings, plant, tools, equipment and site materials safely, and to ensure that sub-contractors do likewise.

As far as the works are concerned, they are unlikely to proceed and clauses 8.12.3 to 8.12.5 require the contractor to prepare an account for submission to the employer within two months of termination and, after taking into account amounts previously paid under the contract, the employer shall pay to the contractor the amount properly due within 28 days of submission. Despite the contract stating that the contractor should be paid within 28 days of submission, contractors are unsecured creditors and, in practical terms, they will join the list of other unsecured creditors waiting hopefully to be paid a percentage of the employer's liquidated value.

The employment of the design team will not necessarily end with the employer's liquidation – their terms of appointment should be checked. However, if they are to continue, they should seek undertakings from the liquidator that they will be paid for their services. If the liquidator is prepared to give such undertakings they can then proceed with their respective duties in the winding down of the contract, the settlement of accounts and the resolution of claims.

In some cases, a funder may exercise 'step-in rights' to take over the contract in which case the employer's debts will be picked up and the contract can continue; this is a subject in itself and beyond the remit of this book.

	£	£
Amount of original contract (with company in liquidation)		200,000
Additions (whether ordered with company in liquidation or completion contractor - priced at rates in the original contract)		10,000
		210,000
Omissions (whether ordered with company in liquidation or completion contractor - priced at rates in the original contract)		7500
Amount of notional final account if original contractor had completed		202,500
Amount of completion contract	30,000	
Additions (for completion contract only - priced at rates in the completion contract)	2500	
	32,500	
Omissions (for completion contract only - priced at rates in the completion contract)	2000	
Amount of final account of completion contract	30,500	
Additional professional fees incurred	1500	
Amount certified and paid to original contractor before liquidation	169,000	201,000
Debt payable by employer to liquidator or receiver		£1500

Example 31.1 Financial statement at completion showing a debt from employer to contractor.

	£	£
Amount of original contract (with company in liquidation)		200,000
Additions (whether ordered with company in liquidation or completion contractor - priced at rates in the original contract)		10,000
		210,000
Omissions (whether ordered with company in liquidation or completion contractor - priced at rates in the original contract)		7500
Amount of notional final account if original contractor had completed		202,500
Amount of completion contract	150,000	
Additions (for completion contract only - priced at rates in the completion contract)	10,500	
	160,500	
Omissions (for completion contract only - priced at rates in the completion contract)	7700	
Amount of final account of completion contract	152,800	
Additional professional fees incurred	7500	
Amount certified and paid to original contractor before liquidation	72,500	232,800
Debt payable by liquidator or receiver and partly by bond holder (if applicable) to employer		£30,300

Example 31.2 Financial statement at completion showing a debt from contractor to employer.

Chapter 32
Capital Allowances

Introduction

Capital allowances are a form of tax relief on commercial property that are among the most significant and yet most frequently underutilised financial incentives in the UK. The origins of the current legislation go back to 1945, and were seen then as a method of modernising the pre-war system of providing tax relief on the depreciation of capital assets. Over the last sixty years the capital allowance system has evolved to the point that many tax theorists believe that the system needs updating. The Treasury has been carrying out a prolonged consultation process under the banner of corporation tax reform; discussion on capital allowances is ongoing. The current Labour Government, however, remains committed to using the tax system to encourage investment in underdeveloped or depressed areas of the country, and so the capital allowances regime continues to be amended on an annual basis as new initiatives are implemented.

Almost all commercial property owners and users who incur capital expenditure on property are entitled to claim capital allowances. As much as 40% of construction costs and up to 75% of fitting-out costs can be allowable. However, research indicates that about half of all businesses miss out on such opportunities. This is partly because the issues are complex often requiring a degree of expertise in both construction costs and tax legislation in order to identify, value and agree capital allowances claims with the Inland Revenue.

The main types of capital allowances are:

- **Plant and machinery allowances** Tax relief is allowed on a wide variety of fixtures as specified in the Capital Allowances Act 2001. See the end of this chapter for some details of what may qualify for relief.
- **Industrial buildings allowances** Relief is given on the entire amount of capital expenditure providing the unit is used for a qualifying trade, as defined in the current Capital Allowances Act.

- **Hotel allowances** Relief is given on the entire capital expenditure providing the property is a qualifying hotel, as set out in the Capital Allowances Act.
- **Research and development allowances** Qualifying trades (generally businesses involved in technological development) can claim for property-related research and development costs.

Eligibility for capital allowances

Claimants must be liable to income tax or corporation tax on profits arising from their business activity. The statute does, however, require further criteria to be fulfilled. To qualify for plant and machinery allowances, tax payers must incur:

- capital expenditure upon the provision of plant and machinery
- that is owned by them and
- is used in the course of their trade.

Claimants tend to be either property investors or occupiers who have incurred capital expenditure on the construction, refurbishment or fitting-out of their premises for the purpose of their business. The basis of any claim is therefore an analysis of actual capital expenditure incurred plus a proportion of associated professional fees.

The worth of capital allowances

The actual cash benefit depends on the tax rate of the business. Every £100,000 of plant and machinery claimed by a business will mean a tax saving of £30,000 over time, assuming a 30% corporation tax rate. Similarly, a higher rate income tax payer would save proportionately more. The total value of plant and machinery is not all claimed in the first year but at a 25% per annum writing down allowance (WDA) claimed on a reducing balance basis. This mechanism is illustrated in Figure 32.1. Since July 1998 the first year's allowance has been available at 40% for some small and medium-sized enterprises, but this reverts to 25% per annum in the second and subsequent years, again on a reducing balance basis. Industrial building allowances and hotel allowances are claimable at 4% per annum from the date of first use until they are exhausted after 25 years. Research and development allowances are all available in the year of expenditure.

Different types of properties will yield different levels of allowances. The following are typical percentages, reflecting the proportion of total construction cost accounted for by plant and machinery:

- Air conditioned offices – 25% to 40%
- Heated offices – 17% to 25%

ASSUMED PLANT AND MACHINERY CLAIM				£500,000
Assuming tax rate @				30%
Calculation of cash benefit				
Year	***Plant and machinery***	***Annual claim (WDA)***	***Cash saving per year @ 30%***	
1	500,000	125,000	**37,500**	
2	375,000	93,750	**28,125**	
3	281,250	70,313	**21,094**	
4	210,938	52,734	**15,820**	
5	158,203	39,551	**11,865**	
6	118,652	29,663	**8,899**	
7	88,989	22,247	**6,674**	
8	66,742	16,685	**5,006**	
9	50,056	12,514	**3,754**	
10	37,542	9,386	**2,816**	
	Total cash saved over 10 years		**£150,000**	
	Allowances go beyond year 10			

Figure 32.1 Cash flow illustration.

- Hotels – 30% to 50% (plus hotel allowances)
- Industrial properties – 5% to 12% (plus industrial building allowances)
- Fitting-out contracts – 40% to 75%.

Tax deductions for property refurbishment schemes

Capital or revenue?

When a business carries out a property refurbishment, the works will generally include an element of alteration and improvement, as well as an element of repair and maintenance. The works of alteration and improvement are capital in nature and the only tax deduction allowed for such expenditure will be capital allowances for qualifying buildings and/or plant and machinery. Repairs and maintenance, on the other hand, are revenue in nature and such expenditure is fully tax allowable in the year of expenditure.

Descriptions contained in contract documentation can be misleading when trying to decide if the work is either a repair or an improvement.

- If the remedial work is related to an alteration or addition to the property, such as a new doorway or window, this is probably capital expenditure.
- If there is any element of improvement or the 'asset' is replaced in its entirety, this is probably capital expenditure.

- If an item is replaced with a modern equivalent, this is likely to be revenue expenditure. (For example, the Inland Revenue will now accept that the replacement of single glazed windows with double glazed windows can be revenue expenditure.)

A much wider understanding of the nature and extent of the refurbishment is necessary to identify correctly the scope and extent of the repair element. The answer to the question will typically be one of fact and degree.

Repair is restoration by renewal or replacement of subsidiary parts of the whole 'asset'. Case law suggests great importance is placed on the principle of partial replacement of an asset. Problems arise in defining exactly what an 'asset' is, because it can be a part of the building, such as an industrial chimney or the roof, as well as the entire building. The contrasting decisions in the cases of *O'Grady* v. *Bullcroft Main Collieries Ltd* (1932) and *Samuel Jones & Co (Devonvale) Ltd* v. *CIR* (1951) serve to illustrate some of the difficulties.

If refurbishment takes place immediately after acquisition, the entire cost will usually be capital expenditure unless it can be shown that the following three principles laid down many years ago in the Odeon Theatres case (1971) have been met:

- the state of disrepair prior to purchase did not 'reduce' the purchase price
- the extent of the repairs required did not affect the use of the property in any way
- trading could have continued without refurbishment.

Any expenditure on a property that could not be occupied or re-let after acquisition without the work in question first being completed will probably be capital in nature.

Incidental capital expenditure

When claiming capital allowances for alterations to existing buildings, one can claim for the expenditure incidental to the installation of any qualifying plant and machinery. For example, the cost of constructing a lift shaft inside an existing building is treated as incidental expenditure to the provision of a lift.

Deferred revenue expenditure

Deferred revenue expenditure is repairs and maintenance expenditure that has been capitalised, instead of writing it off immediately to the profit and loss account as it is incurred.

Until recently, it was common for companies to capitalise the full cost of a refurbishment project in their annual accounts and then later, when the tax computations were prepared, claim a full deduction for the items of a revenue nature.

This approach meant that it was possible to maximise tax relief, but at the same time minimise the impact on reported profits.

The Inland Revenue has now stated in *Tax Bulletin (Issue 53)* that deferred revenue expenditure should only be allowed against tax when it is charged to the profit and loss account in accordance with generally accepted accountancy practice. This means that deferred revenue expenditure will only be allowed in accordance with a company's depreciation policy. The Inland Revenue has applied this approach to tax computations for accounting periods starting after 30 June 1999.

For example, a property company incurs £300,000 on a refurbishment project of which £50,000 relates to revenue expenditure. The company capitalises all £300,000 in the accounts. The property is held on a 50 year lease and the depreciation policy of the company is to write off the leasehold property at 2% per annum over the remaining life of the lease. In this case, the tax deduction for deferred revenue expenditure would be £1000 per annum (£50,000 × 2%) rather than a full deduction of £50,000 in the year of expenditure. Of course, one might say that property companies do not usually depreciate their properties. In this case, no tax deduction would be available for deferred revenue expenditure until the property is sold.

It is perhaps worth considering the capital gains tax position when there is deferred revenue expenditure. Using the previous example as an illustration, say the property was sold in year three – the base cost for the refurbishment expenditure would be £250,000 as £50,000 was subsequently identified as revenue expenditure, notwithstanding that the total of £300,000 was capitalised in the accounts.

In both year one and year two £1000 each would have been allowed against tax as revenue expenditure. On the sale of the property the remaining £48,000 of deferred revenue expenditure would also be allowed. Therefore, the full cost of £300,000 is eventually allowed against tax when the property is sold.

In conclusion, if property companies desire to minimise their tax burden, then a detailed analysis of their annual refurbishment expenditure should be undertaken prior to the preparation of their annual accounts. This will enable the identification of revenue expenditure in the accounts, with a full deduction against tax being obtained in the year of expenditure.

Action summary

In order to ensure the correct tax treatment and maximise the deductions available, the following are suggested as some of the key actions:

- Obtain drawings showing the building before the works commenced and after the works are completed.
- Consider whether an entire asset has been wholly replaced or merely a subsidiary part.
- With recently purchased property apply the Odeon Theatre case tests.

- Consider the primary purpose of the works as a whole – are the works really so extensive that the project is more one of redevelopment, rather than repair and refurbishment?

From a tax perspective, expenditure should first be allocated as repair and maintenance revenue expenditure giving rise to an immediate 100% deduction, second as capital expenditure qualifying for plant and machinery allowances at the rate of 25% per year (or a 50% or 40% first year allowance if the business is a small or medium sized enterprise) and third, if applicable, as capital expenditure qualifying for industrial building (or hotel) allowances at the rate of 4% per year, with the remainder not qualifying for any deduction.

Energy saving plant and machinery

The Enhanced Capital Allowances (ECAs) scheme for energy saving plant and machinery came into effect on 1 April 2001. Under the scheme businesses are able to claim 100% first year allowances on investments in energy saving plant and machinery. Leased assets only qualify from 17 April 2002.

The legislation was included in the Finance Act 2001 as Section 65 and Schedule 17 and revises the Capital Allowances Act 2001. A Treasury Order (SI2541) giving full effect to the Energy Technology Criteria List and the Energy Technology Product List referred to below came into force on 7 August 2001.

Key features

At the beginning of 2004 there were 13 energy saving technologies that met the relevant energy saving criteria:

- combined heat and power (CHP)
- boilers
- motors
- variable speed drives
- refrigeration equipment
- pipe insulation
- thermal screens
- lighting
- heat pumps for space heating
- warm air and radiant heaters
- solar thermal systems
- automatic monitoring and targeting equipment and
- compressed air equipment.

Qualifying products within these technology categories appear on the Energy Technology Product List, except for pipe insulation, lighting and CHP, which must satisfy the criteria specified in the Energy Technology Criteria List.

A further three technologies (air to air energy recovery, compact heat exchangers and HVAC zone controls) were added in Summer 2004. Various amendments and additions were also made to the existing technology categories.

The Finance Act 2003 introduced a new category of environmentally beneficial plant and machinery qualifying for 100% first year allowances. The Water Technology List includes five categories of sustainable water use products, namely water meters, flow controllers, leak detection equipment, low flush toilets and efficient taps. Rainwater harvesting equipment was added to the list in late 2004.

Key features of the scheme are:

- All businesses will be able to claim ECAs, regardless of size, industrial or commercial sector or location.
- Only investments in new and unused plant and machinery can qualify.
- ECAs will permit the full cost of the investment to be relieved for tax purposes against taxable income of the period of investment.
- Qualifying technologies will have to meet defined energy saving criteria published in a list.
- There will be an annual review of technologies and criteria.
- There are no territorial restrictions on manufacturers wishing to place their products on the list, or the source of products.

Key exclusions

There are some major types of expenditure excluded that remove a substantial amount of investment from potentially qualifying under the scheme:

- Long life assets, that is an asset with an expected useful economic life of 25 years or more when new.
- Assets that are buildings or structures as defined in the Capital Allowances Act 2001, Section 21 List A and Section 22 List B, as these assets do not qualify for any Capital Allowances given on plant and machinery.

The second of these exclusions left a particular question mark against the extent to which one of the technologies, lighting, qualified under the scheme. Section 23 List C removes lighting systems from the exclusion referred to above. However, there are caveats on the exception and the Inland Revenue will only accept that lighting qualifies in limited circumstances, irrespective of whether it is energy saving or otherwise.

How to claim ECAs

ECAs are claimed in the same way as other Capital Allowances, by inclusion in the company, individual or partnership tax return.

If a product on the Energy Technology Product List is purchased the claim will be based on the price paid for the item. In addition, the tax payer may claim on any costs that are directly associated with the provision of the product. The ECA website suggests these may include:

- transportation and installation costs including, for example, project management costs and commissioning
- professional fees directly related to the acquisition and installation of the asset
- costs of alteration to an existing building incidental to the installation of the asset.

When a larger item of plant and machinery is acquired that has a qualifying product already installed in it as a component, then the proportion of the cost that relates to the qualifying product will qualify for an ECA. The remainder of the equipment will attract allowances at the normal rate. The government specifies the amount that can be claimed for a qualifying product incorporated into another piece of equipment – details can be found on the relevant website.

Qualifying products and items

Qualifying products within the technology categories appear on the Product List. Qualifying items within the pipework insulation, lighting and CHP categories do not and instead are covered by the Energy Technology Criteria List.

Lighting must be energy saving and is divided into three categories:

- lamps that can be substituted for existing less efficient types
- lighting fittings complete with required lamps and control gear
- lighting control equipment.

Performance benchmarks in terms of energy efficiency requirements have been defined for all product descriptions. Manufacturers, suppliers and distributors are directed to provide customers with advice as to whether their products qualify. Manufacturers are encouraged to adopt and label their products with lighting fitting efficiency codes and the ECA website also includes a system for overall installation energy appraisal.

CHP is the simultaneous generation of heat and power (usually electricity) in a single process. Heat that would otherwise be wasted is supplied for use in industry or in community heating. ECAs may be claimed on plant and machinery in a good quality CHP installation, provided that the main intended business will be to provide heat and power for clearly identified users, either on site or known third parties.

In order to qualify for ECAs each CHP scheme must be self-assessed and validated under the CHP Quality Assurance Programme (CHPQA). The applications should be made once detailed information is available, but before significant expenditure is incurred. For more information visit the DETR's official website at www.eca.gov.uk.

Pre-contract planning to optimise capital allowances

Pre-contract planning is a pre-requisite to maximising a tax payer's entitlement to capital allowances. The aim of the exercise is to compile documentary evidence and use precedents in case law to strengthen and support the claim. By following the design and construction processes and recording why particular design decisions are made the business use test can be overcome and the value of a claim can be increased by an average of 20%–25% on prestige and other complex new buildings, refurbishment and fitting-out contracts.

It is not always the nature of the item itself that determines eligibility for capital allowances. The way particular features or building elements function in the trade of the end user will govern whether an item passes or fails the business use test. The following list is typical of the items on which a tax planning exercise would focus:

- builders' work in connection with services
- powered doors or shutters
- built-in furniture
- cold water services installation
- above-ground drainage installation
- electrical installation
- raised demountable floors
- suspended ceilings (other than plenum ceilings)
- fire safety requirements in existing buildings
- demountable partitions
- preliminaries, professional fees and other project 'on-costs'.

Some questions and answers

Can I get capital allowances for any of my fitting-out works?
Yes. There could be a significant level of capital allowances available as a result of incurring capital expenditure on fitting-out works. It is not uncommon to find that as much as 75% of such expenditure qualifies for plant and machinery allowances.

What is plant and machinery?
The Capital Allowances Act precludes most parts of buildings and permanent structures from qualifying for capital allowances as plant and machinery. However items that could qualify include air conditioning, heating, lifts, electrical installations (either in part or in their entirety, depending on the circumstances), sanitary ware, hot water, carpets, blinds, demountable partitioning, signage, and certain types of access flooring systems and suspended ceilings. In retail stores, branch offices and similar premises, large sums of money are often spent on achieving the corporate image. In hotels, public houses and restaurants, money is spent on decorative assets to create the right atmosphere or ambience. Where it can be shown that these items pass what the courts have described as the 'business use' test and the 'premises' test, the expenditure on them should qualify for capital allowances. It is therefore important to bear in mind that it is not just the item but also the reason why it was installed that is important.

I do not have a breakdown of the fitting-out costs; can I still claim?
Yes. Problems often occur with lump-sum contracts priced on the basis of a specification and drawings. If, however, the total cost is broken down using traditional surveying skills and then documented, a claim can be made on the identified qualifying expenditure.

Does self-assessment affect capital allowances?
Yes. It affects the way claims are prepared. Previously a small minority often submitted percentage claims with little or no back-up in support of the figures. Self-assessment has introduced new obligations for both tax payers and advisers with penalties for non-compliance. This means being able to produce a complete audit trail whether or not the Inland Revenue actually calls for it.

What about capital allowances on refurbishment schemes?
Capital expenditure on these types of projects will attract high levels of capital allowances. The key area to consider is works incidental to the installation of plant and machinery. The Capital Allowances Act allows expenditure on alterations to existing buildings which are incidental to the installation of plant and machinery to be treated as part of the qualifying expenditure. In addition, some of the expenditure may be reclassified as a revenue expense, the whole of which would, in consequence, be deductible in the year in which it was incurred.

Does the building design affect the level of capital allowances?
Yes. The more complex a building design, the greater the scope for tax planning capital allowances. Specially designed, newly constructed, owner-occupied properties can qualify for capital allowances as high as 40% of total expenditure. Although it may not always make long-term economic sense to choose design and engineering solutions purely because they attract capital allowances, it is worth commissioning a design 'tax' audit. Then, when decisions are made, the tax consequences are known. On projects worth many millions of pounds, even small changes, initiated at an early enough stage, can have a major influence on the ultimate eligibility of items for capital allowances.

Index

Page numbers in *italics* refer to figures